Computational Molecular Science and Engineering Forum 2019

AA009222

Held at the 2019 AIChE Annual Meeting

Orlando, Florida, USA
10 - 15 November 2019

ISBN: 978-1-7138-0542-7

Computational Molecular Science and Engineering Forum 2019

Held at the 2019 AIChE Annual Meeting

Orlando, Florida, USA
10 - 15 November 2019

AIChE
The Global Home of Chemical Engineers

Printed from e-media with permission by:

Curran Associates, Inc.
57 Morehouse Lane
Red Hook, NY 12571

Some format issues inherent in the e-media version may also appear in this print version.

Copyright© (2019) by AIChE
All rights reserved.

Printed with permission by Curran Associates, Inc. (2020)

For permission requests, please contact AIChE
at the address below.

AIChE
120 Wall Street, FL 23
New York, NY 10005-4020

Phone: (800) 242-4363
Fax: (203) 775-5177

www.aiche.org

Additional copies of this publication are available from:

Curran Associates, Inc.
57 Morehouse Lane
Red Hook, NY 12571 USA
Phone: 845-758-0400
Fax: 845-758-2633
Email: curran@proceedings.com
Web: www.proceedings.com

TABLE OF CONTENTS

569983 Chemical and Mechanical Stability of a Model Ionic Liquid Epoxy Material Exposed to Galactic Cosmic Ray in a Simulated Martian Environment

With NASA's "Journey to Mars" mission progressing at a rapid rate, development of novel space materials that would withstand the extreme Martian environment has galvanized the materials research community. Ionic liquids (ILs) are currently being investigated as potential binders for Martian regolith in NASA's *in situ* resource utilization efforts for the construction of Mars habitats. Ionic liquid epoxy (ILE) systems as potential IL derivatives also possess promising properties that may justify their use in space composite materials. To obtain a fundamental understanding of the chemical and mechanical stability of a model ILE system exposed to the extreme Martian environment, i.e., exposure to Galactic Cosmic Ray (GCR) and cold temperature extremes, we performed a series of molecular dynamics (MD) simulations, wherein a model amorphous ILE system composed of cationic imidazolium bis(epoxide) monomers with their bis(trifluoromethane)sulfonimide ($[Tf2N]^-$) anionic pairs were first energy-minimized and equilibrated at room temperature in a 3D-periodic system using the Reactive Force Field (ReaxFF). Since, to the best of our knowledge, the ReaxFF potential has not been used for an ILE system in the past, we parameterized the potential to obtain a stable ILE system and validated it by comparing the simulated ILE physical properties to those obtained experimentally. Next, the system was subjected to high-velocity inert He atoms, with their kinetic energy rate equivalent to the reported GCR energy rate of 30 μS/h on the surface of Mars, in a 2D-periodic system with a vacuum slab on top. The system temperature in the 2D-periodic system was first equilibrated at 210.15 K (-63°C), which is the average surface temperature on Mars near the Viking 1 Lander site. To account for statistical variations in the data, we used a total of three replicates in our simulations and used the average values for the calculated properties during post-processing. Finally, we determined the average mass loss, chemical species evolution, damage propagation depth, as well as initial/final stress-strain responses of the replicate material systems to evaluate the chemical and mechanical stability of the model ILE, respectively.

Copyright © American Institute of Chemical Engineers. All rights reserved.

573597 An Intermittent QM/MM Technique to Model Force Induced Bond Scission in Polymeric Systems

The materials in our bodies are constantly experiencing and responding to external forces on scales both large (*e.g.* forces impacted on joints and ligaments during exercise) and small (*e.g.* hydrodynamic shear experienced by proteins and cells). Under such conditions, mechano-chemistry – chemical changes induced by mechanical stimuli – can play a large role in a number of biological processes. These include both events that modify biological function, such as the redox switching behavior of disulfide bonds and instances where a biomolecule assumes a more passive role as a structural material such as actin filaments in muscle or collagen fibers in tendons and other load bearing tissue. Currently, modeling of these single-bond events usually present significant time and size limitations and it is therefore difficult to capture the effect that these small-scale phenomena impart on the large-scale behavior of the material. Modeling these bond-breaking phenomena in the context of their surrounding biomolecular environment has not yet been accomplished, and could allow for the observation of many physiological mechanisms and further exploration of the mechano-chemistry of proteins and protein function. To this end, we have developed a computational workflow that attempts to bridge this gap by using classical molecular dynamics (MD) simulations that are intermittently enhanced on an *ad-hoc* basis with local quantum calculations using Density Functional based Tight Binding (DFTB). This is accomplished by applying force/ distance cut-offs on load-bearing bonds which trigger the QM/MM calculations to determine whether or not bond scission occurs before switching back to the more computationally efficient MD framework. We demonstrate success on model systems of collagen and sulfide-bond containing proteins and compare these results with experimental measurements. Further refinement and automation of this technique will allow for increases in scale with the goal of creating a multidisciplinary implementation capable of allowing the investigation of problems in fields such as materials science, engineering, and biology.

Copyright © American Institute of Chemical Engineers. All rights reserved.

577992 Study of Orientational Dynamics in a Cross-Linked Epoxy Network Using Atomistic Simulations

Subtle differences in the chemistry of cross-linked epoxy networks can lead to substantial changes in their thermomechanical properties, indicating the strong connection between chemical architecture and polymer physics in these materials. Atomistic molecular dynamics (MD) simulations leveraged with time-temperature superposition (TTS) are well-suited for studying these systems, as the combination enables integration of molecular length and time scale insight with experimental measurement at times scales relevant to commercial applications and measurement science. In this work, we study the orientational dynamics of cross-linked epoxy using the combination of atomistic MD and TTS. Specifically, we characterize the rotational diffusion autocorrelation function for network features at both atomic and molecular length scales at temperatures spanning from the rubbery state across the glass transition. At atomic length scales, we study the bond vectors of the central carbon atom of the epoxy monomer and the nitrogen atom of the cross-linker. At molecular length scales, we study end-to-end vectors of the epoxy monomer and the cross-linker. Using TTS, we show that all the molecular level data can be collapsed onto master curves spanning over nine orders of magnitude in reduced time. The temporal features of master curves can be quantitatively compared with the trend in experimental creep compliance for the same system. At the atomic level, TTS shows that the topological location of the atoms results in distinctly different signatures in the orientational dynamics for the two bond vectors. The results are relevant to the behavior of the molecular probes (mechanophores) being used for damage metrology by our experimental collaborators.

Copyright © American Institute of Chemical Engineers. All rights reserved.

573451 Machine-Learning on the Coarse-Grained Polymer Genome

The recent emergence of machine-learning (ML) techniques in tandem with molecular modeling has greatly enhanced simulation capabilities. The possibility of using existing and/or generated data within ML frameworks to greatly accelerate property prediction is particularly intriguing for materials design. While there has been significant advancement and progress in the context of small molecules or crystal structures, this objective is notably difficult for polymers or soft materials, for which target properties are often characterized by distributions resulting from an ensemble of configurations governed by subtle interactions. In this talk, I will discuss the viability of using coarse-grained simulations in tandem with supervised machine-learning to predict polymer properties based on polymer sequence, i.e., to define the chemistry/structure->property relationships that characterize the polymer genome. Within this framework, we systematically interrogate the influence polymer composition and topology on observed polymer properties for a defined coarse-grained chemical space. I will further show how this framework lends itself to generative design strategies to achieve target properties. Overall, this work aims to highlight the promising capabilities (and difficulties) in using ML towards soft materials design.

Copyright © American Institute of Chemical Engineers. All rights reserved.

570627 Biased Sampling of 3D Polymer Conformations in an External Field Using Brownian Bridges

The equilibrium conformation of polymer molecules in an external field is of fundamental interest to the calculation of macroscopic polymer properties. Here, we build upon a mathematical method [Krishnaswami et al. (1997). J. Chem. Phys., 107(15), 5929-5944.] for efficient sampling of continuous polymer chains described by a stochastic differential equation along the backbone's contour length (e.g., Gaussian chains, worm-like chains, etc.). This method is based on the concept of a Brownian bridge, which is a stochastic process whose endpoint is constrained through the addition of a biasing drift derived from Bayes theorem. We note for a polymer system, one can formulate a Brownian bridge to exclude small Boltzmann weights during sampling, and hence only sample regions of high probability in phase space. In this talk, we formulate numerical schemes to generate such bridges in applications where one wants to sample polymer chains in an arbitrary, external potential. The biasing drift needed for the Brownian bridge is a hitting probability (which can be thought of as an entropic potential), which we obtain by solving a Backwards Kolmogorov equation (or equivalently, a Backwards Fokker Planck equation). Thus, through an additional cost of solving a PDE once at the beginning of a simulation, we can guarantee our stochastic differential equations will be able to generate efficient configurations of polymer chains. We show examples for a wide range of potentials where polymer conformations are shown to satisfy exact energy constraints either at a fixed value or within a stipulated range. We conclude by discussing how to extend this technique to include segment-segment interactions (e.g., excluded volume) and multiple chains. We also discuss how to obtain approximate biasing, entropic potentials through approximations to the Backwards Kolmogorov equation for more complicated situations.

Copyright © American Institute of Chemical Engineers. All rights reserved.

574670 Developing a Tractable Simulation of Primary Crystal Nucleation in a Polymer Melt

While semi-crystalline polymers are ubiquitous and have been studied for decades, many mysteries remain regarding the kinetics of crystallization, limiting our control of material properties and impeding future technological innovations. One especially perplexing phenomena is the so-called melt memory effect in polymer re-crystallization experiments where properties such as the nucleation rate and crystal thickness are observed to depend strongly on process history. It is widely suspected that melt memory effects are tied to the primary nucleation event in polymer crystallization, however the community is divided between two fundamental explanations. Some researchers believe that slow dynamics, primarily due to entanglement, are the root cause of memory effects in what would otherwise be a straightforward application of classical nucleation theory. However, driven by a number of intriguing theoretical and experimental results, some researchers have suggested that melt memory effects are better explained by a more sophisticated theory of primary nucleation that includes one or more intermediate metastable states. To explore the latter hypothesis, we are developing a Monte Carlo model capable of calculating the free energy surface for a nucleating crystal in a polymer melt. We will discuss our recent progress and challenges developing this model, focusing on our efforts to create a scalable, parallel Wang-Landau simulation of a polymer melt.

Copyright © American Institute of Chemical Engineers. All rights reserved.

575510 Simulation of Peg/Poly Aspartic Acid Micelles Provides Insight for Design of Drug-Delivery Systems

We use an on-the-fly biasing method to explore the conformational profile of poly(ethylene glycol)/poly(aspartic acid) block copolymer molecules in lipozomal micelles. The configuration of assembled lipozome micelles is of importance when designing micelle-based small drug molecule delivery systems. However, these configurations are difficult to measure experimentally, and the timescales required for micelle assembly in-silico can be prohibitive as well. In this work, we use a custom plugin written in our group, HOOMD-TF, that allows GPU-accelerated machine learning via Google's TensorFlow library in tandem with the HOOMD-blue molecular dynamics simulation engine to simulate a coarse-grained poly(ethylene glycol) and poly(aspartic acid) block copolymer micelle in water. The polymer concentration is 1.25mg/mL, with a size of 117 ethylene glycol and 48 aspartic acid residues, and is simulated in the NVT ensemble at room temperature. We use HOOMD-TF to employ an on-the-fly biasing method to match the system configuration to experimental X-ray scattering data and investigate the conformational space of these micelles. We perform cluster analysis of the polymer backbone angles to find the most common configurations, and sample the free energy surface along angular space and radius of gyration. We also measure the observed micelle internal and external diameters and compare with experimental observations. This information can be used to help select suitable small drug molecules for encapsulation, and to design future polymer micelle drug-delivery systems.

Copyright © American Institute of Chemical Engineers. All rights reserved.

565498 Prediction of Self-Diffusion Coefficients of Polymer Melts Using Modified Free Volume Theory and MD Simulation

MD simulation of polyethylene melts with different chain lengths is performed. Crossover in the center-of-mass self-diffusion coefficient (D) as a function of chain length (N) is observed: as $N<N_c$, $D \sim N^{-1}$ and as $N>N_c$, $D \sim N^{-2.2}$. The modified free volume theory is used to describe such crossover as observed in the MD simulation. The two parameters in the modified free volume theory, which are the critical free volume and mean free volume per repeat unit, can be determined from the pair correlation function by either solving PRISM equations for polymers or using the MD simulation data. This makes the modified free volume theory parameter-free. Voronoi tessellation of the MD simulation data are also performed for comparison with the statistical mechanical calculations. Such analysis gives us the exact distribution of free volume in the polymer melts, and therefore the probability for a repeat unit having free volume greater than that of critical free volume.

Copyright © American Institute of Chemical Engineers. All rights reserved.

556963 Deep Learning of Activation Energies and Automated Reaction Dataset Generation

Automated mechanism generation is an efficient tool for rapidly constructing large-scale chemical reaction mechanisms, which are necessary for realistic modeling of complex processes, such as combustion and oxidation, soot formation, or pyrolysis. Nonetheless, hard-coded reaction templates employed by mechanism generation software do not cover the set of all possible reactions that might be important. Moreover, only a limited quantity of high-quality data is available for the estimation of kinetic rates during mechanism generation, which may result in poor estimates and the erroneous exclusion of important reactions. In response, we have developed methods to automatically generate a large database of chemical reactions and designed a template-free deep learning model capable of providing a quick assessment of potentially important reactions that are missing from a reaction mechanism.

We use activation energy as a temperature-independent measure to gauge the potential importance of new reactions and train a machine learning model to estimate the energies by relying on a reaction encoding which predominantly focuses on the atoms involved in the reactive center. In order to predict kinetically relevant activation energies as well as being able to tell which reactions have barriers that are so large that they cannot possibly be relevant, the training data for the machine learning model must span a more diverse set of reactions than is currently available in the literature. We use an automated transition state finding algorithm to generate a diverse training set of tens of thousands of reactions, and thus create an activation energy prediction model that can quantify the importance of proposed reactions.

Copyright © American Institute of Chemical Engineers. All rights reserved.

566431 Towards Automated Discovery of Plausible Reaction Paths in Complex Catalytic Systems Using Network Generation and Optimization

The application of state-of-the-art computational tools, viz. density functional theory (DFT) to calculate the energetics of every species and transition state and microkinetic modeling to obtain reaction rates and yields is computationally intractable for complex reaction networks comprising of thousands of reactions and intermediates. In this talk, we present a computationally tractable method to identify dominant reaction pathways for such complex heterogenous catalytic reaction systems. The method: (i) constructs the reaction network in an automated manner using the knowledge of potential elementary steps, (ii) uses simple data-driven models derived from DFT calculations that can then be applied to the rest of the network, (iii) employs a mixed-integer optimization problem to "screen" for the dominant pathways based on energetic feasibility metrics or maximum reaction rate estimates, (iv) rank orders potential alternative solutions of the optimal pathways within the errors of density functional theory and the data-driven model, and (v) generates volcano plots based on known linear scaling correlations for catalyst screening.

The details of the proposed method will be discussed in the context of demonstrative examples comprising of oxygenate decomposition and dehydrogenation of organic hydrogen carriers on transition metal catalysts.

Copyright © American Institute of Chemical Engineers. All rights reserved.

572387 Robust Catalyst Characterization and Estimation of Mechanistic Pathways through Data Science

Robust Catalyst Characterization and Estimation of Mechanistic Pathways through Data Science

Transient kinetic experiments provide a significant volume of data as the material is evaluated as a function of time rather than at a singular steady-state. This allows more detailed microkinetic information to be collected where variations in temperature, pressure or concentration on the materials surface is reflected in the measured response. However, the resolution of this detail also requires the development of new analysis to accurately inform the user of the underlying mechanism while maintaining the temporal nature of the data. A novel approach presented here utilizes a combination of modern machine learning techniques and physics-based assumptions to analyze the full information available from transient kinetic experiments without assumption of a kinetic model. Furthermore, this approach applies feature selection within a global reaction network to provide an optimal trade-off between the selection of a complex estimated mechanism and the minimization of the fit on the measured reaction rate. As an example, a set of catalysts are compared when performing ammonia decomposition in transient pulse response experiments. We were able to quantitatively measure the ammonia gas interaction with different surface species and the degree of impact on the transformation rate. As such, the global sparse covariance matrix describes the network path of individual catalysts. More specifically, we are able to distinguish differences in each catalyst based on the degree, if at all, of the complete set of surface and gas phase interaction combinations. The figure below shows an example of the obtained network where the surface species are indicated as z and the gas species are indicated as c. The coloration of each node indicates the grouping of the interactions while the edges indicate a significant interaction. These connections between the gas and surface species are remarkably distinct for different materials and represent a mechanistic signature for discriminating catalysts. The results of this novel method provide interpretability through integration of physical assumptions, estimation of the mechanistic reaction pathways and a unique experimental kinetic fingerprint for each material whereby catalysts can be microkinetically discriminated.

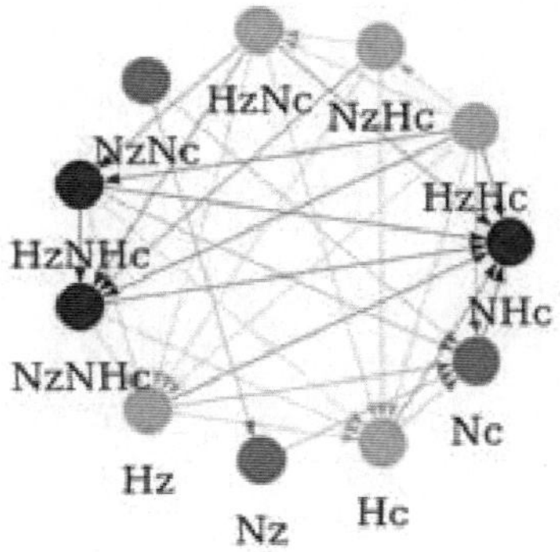

Copyright © American Institute of Chemical Engineers. All rights reserved.

573166 Autoneb: A Toolkit for Automating Reaction Network Calculation

We present an automated toolkit for the first-principles studies of reaction pathways, especially in complicated reaction networks involving multiple conformers of the reaction species. Challenges remain in the automatic generation of conformers of an arbitrary molecule due to the high variability in bonding and charge state and the limited availability of suitable force field parameters needed in conventional conformer search methods, especially for charged species or radicals. We overcome this through automated enumeration of all possible combination of rotatable dihedrals, followed by automated pruning to generate featured conformers. Pruning of the conformers is achieved through a 3-step process: (1) elimination of conformers with atom clashes based on molecular fingerprint, (2) geometry optimization with density functional theory and elimination of energetically unfavorable conformers, (3) extraction of featured conformers via a clustering algorithm based on the geometries. For the surviving conformers, nudged elastic band (NEB) calculations are then automatically set up for each possible reactant-product conformer pair in the reaction network. A new interpolation scheme is implemented in AutoNEB for generating the initial structures of the "beads" in NEB, in order to properly handle reactions involving complicated motion in the internal coordinates. Finally, automated post-processing aids the discovery of the most favorable route in the whole network, with visualization of the energy profile. This toolkit has been applied to the reaction network for gas-phase Pomeranz-Fritsch synthesis of isoquinoline and has provided an insightful explanation of the phenomena observed in mass spectroscopy.

Copyright © American Institute of Chemical Engineers. All rights reserved.

575887 Recent Advances in Automated Transition State Theory Calculations: A Case Study Using Butanol

With the effects of global climate change becoming more apparent, researchers have been pushing for a shift from conventional fossil fuels to greener alternatives. However, these fuels need to be adequately studied *a priori* to make reasonable experimental choices. In these studies, detailed reaction mechanisms need to be constructed to explain the complex combustion phenomena. Because of the sheer number of thermodynamic and kinetic parameters, they are often estimated rather than determined experimentally. *Ab initio* calculations and transition state theory (TST) can determine thermodynamic and kinetic parameters more accurately than commonly used estimates. However, given the number of reactions and species present in even simple combustion systems, these calculations need to be automated.

Many researchers have been working on tools to perform these automated TST calculations, but for this work we aim to improve AutoTST, the automated transition state theory code first developed by Bhoorasingh et al. [1]. The first generation of AutoTST would start by creating a transition state geometry guess by using a functional group tree to decide on key distances in the reaction center and using RDKit [2], an open source cheminformatics toolkit, to generate the starting geometry. AutoTST would then employ Gaussian [3] to perform a series of partial optimizations to arrive at an optimized transition state complex. This complex is then validated through automated intrinsic reaction coordinate (IRC) calculations and validated geometries are then fed into Arkane [4] to obtain kinetic parameters. This work has improved AutoTST by: rewriting AutoTST as an independent, reusable, module; including the effects of 1-D hindered rotors; adding a systematic conformer analysis; and bypassing the expensive IRC calculations by using vibrational analysis to validate transition states.

To assess the impact of these improvements, we performed automated TST calculations on 1117 reactions and 298 species present in a model for the combustion of butanol [6] and compared these reaction rates and thermodynamic expressions against values calculated by the first generation of AutoTST, values estimated by RMG, and values published in the model. From here, we identified reactions and species where our calculations deviated the most from comparisons, perform benchmark calculations to generate high fidelity kinetics and thermodynamic parameters for these reactions and species, and compare and investigate the sources of these discrepancies. This framework will enable researchers the ability to generate large amounts of high fidelity kinetic and thermodynamic training data on-the-fly for machine-learning based estimation methods.

References:

[1] Bhoorasingh, P. L.; Slakman, B. L.; Khanshan, F. S.; Cain, J. Y.; West, R. H. Automated Transition State Theory Calculations for High-Throughput Kinetics. J. of Phys. Chem. A. 2017, 121, 6896-6904.

[2] RDKit: Open-source cheminformatics; http://www.rdkit.org

[3] Gaussian 16, Revision B.01, Frisch M. J. et al., Gaussian, Inc., Wallingford CT, 2016.

[4] Allen, J. W.; Green, W. H.; Arkane: http://reactionmechanismgenerator.github.io/RMG-Py/users/arkane/credits.html

[5] Gao, C. W.; Allen, J. W.; Green, W. H.; West, R. H. Reaction Mechanism Generator: Automatic Construction of Chemical Kinetic Mechanisms. Comput. Phys. Commun. 2016, 203, 212–225.

[6] Sarathy S. M. et al. A comprehensive chemical kinetic combustion model for the four butanol isomers. Comb. Flame. 2016, 159, 2028-2055.

Copyright © American Institute of Chemical Engineers. All rights reserved.

577241 Smart Synthetic Path Generation System with Analysis of Reaction Risk Enhanced By Automated Open Data Collection

Chemical accidents often occur during laboratory experiments, pilot plant operations and abnormal reactor operations. To prevent these incidents, it is necessary to find relevant information before starting the synthesis experiment as well as process operations. In the process design stage, reaction information is essential to prevent reaction runaway. To get information, there are various sources of information related to synthesis, including the Internet and literature, and it takes a long time to search. In order to solve this problem, we propose an intelligent support system for synthetic path generation.

The DB used in the system is constructed by collecting open data about chemicals, synthesis, and safety (e.g., NFPA 704, GHS, MSDS, PAC, etc.). The data about chemical substances are available in freely accessible chemical DBs, which can be found in PubChem, ChEMBL, etc., and synthesis data can also be found through various sources. However, it is difficult to organize and search through this various set of data. For this reason, Web Scraping and Web Crawling techniques based on Python are used to explore, collect, organize, and obtain data from diverse sources. Since safety information offered by each Web site is different, safety data for chemicals are collected separately and made into a unified DB. The path generation algorithm is a combination of Depth-First Search and pruning algorithm. The suggested algorithm proceeds with the search based on the target substance and cuts off the child nodes if it is judged that there is no possibility for the child nodes. As a result, only the nodes that have possibility of synthesis are left, and paths are generated by connecting the nodes.

The results of generation show a synthetic path graph of various steps and information of each path for the target substance. This graph shows possible synthetic paths connecting from the target substance node to the starting material node and the user can easily assess the whole path. Synthesis data nodes containing hazardous chemicals are graded and displayed in a separate color (green, yellow or red depending on the severity). Each path includes data such as operating conditions and chemicals used for each synthesis method on the path. For each path where hazardous chemicals are used, additional information is displayed by analyzing the risk in the aspect of safety, health, and environment. After analyzing the synthesis data, the risk class of each synthetic path is classified and weighted for ranking.

The data in the built-in DB are continuously collected and automatically updated. First, about 100,000 synthesis data were obtained and the system test was conducted. We used Scifinder and SciPlanner of CAS to verify the data obtained from the test. Scifinder has a large DB based on many literatures and patents, so it is easy to search for synthetic information. As a result of using SciPlanner, there are few cases where there is no result corresponding to the proposed system's results, but most of matching results could be found. It verifies that actually available paths are all generated using the system even if the size of DB becomes large. The system is being released as open source for free use. For the benefit of each research institution, researchers can register their private data and expand the DB according to the format type. It increases the likelihood of getting results that are tailored to their goals. Using the proposed system, it is expected that researchers will find a safer, reliable, and economical reaction pathway by referring to the generated results and help prevent accidents in experiments and eventual plant operations.

Copyright © American Institute of Chemical Engineers. All rights reserved.

572321 Reaction Network Analysis and Motifs for Rhythmic Dynamics in Metabolic and Circadian Models of Cyanobacteria in a Photobioreactor

Reaction network theories are tools for stability analysis of open reacting systems provided that stoichiometric (chemical) equations are given for each reaction step together with power law rate expressions. Based on stoichiometry alone, elementary subnetworks (known also as elementary modes or extreme currents) are identified and their capacity for displaying dynamical instabilities, such as bistability and oscillations, is evaluated by examining the associated Jacobian matrix [1]. This analysis is qualitative in the sense that only reaction orders are needed as input information, whereas rate coefficients and concentrations of chemical components may remain unspecified. In the next step, the subnetworks are combined to form the entire network and its stability is deduced from the stability of the constituting subnetworks. This combination principle can be conveniently used for kinetic parameter estimation of the unknown/unspecified rate coefficients by applying linear optimization to a set of constraint equations balancing linearly combined subnetworks with the corresponding rate expressions [2]. From mathematical point of view, this is a special case of convex optimization.

From the application point of view, the outlined theory is applied to experimentally measured activity of photosynthesizing diazotrophic cyanobacteria in a photobioreactor [3]. In particular, we focus on biochemical processes giving rise to experimentally observed change from a steady state to oscillatory dynamics. The oscillations in cyanobacteria come from either circadian cycles synchronized with external light/dark cycle or from an internal ultradian clock [4], which is active even in the absence of external environmental cues. For the former, we examine models of circadian clock associated with a network involving the KaiABC proteins and their regeneration via a transcriptional network, for the latter, a carbon-nitrogen metabolic model is analyzed.

For these biochemical systems a reaction mechanism is assumed (or available from previous research) with only a limited set of known kinetic parameters, in addition to input/output parameters known from the experiment. Then the set of unknown kinetic parameters is estimated via linear optimization so that the dynamics displayed by the model coincides with the experimentally observed behavior (emergence of oscillations). Such an estimate may not yield the ultimate best fit, rather, it helps to locate a region in the parameter space, where the observed dynamics are reproduced by the model. This global approach is useful especially when the number of unknown parameters is large. Once a suitable parameter region is found, either standard least-square methods or other more refined algorithms involving deep learning may be used to fine tune the parameters values.

In addition, the reaction network theory is useful in identifying subnetworks that are at the core of destabilizing the steady state. In particular, such subnetworks provide prototypes/motifs of (bio)chemical oscillators, which possess a characteristic network topology [5]. Thus the search for dynamical instabilities in large networks typical for biosystems can be reduced to the search for motifs.

References:

[1] B. L. Clarke, Stability of complex reaction networks. Advances in Chemical Physics 43, 1-278 (1980).

[2] V. Radojkovic and I. Schreiber, Constrained Stoichiometric Network Analysis, Physical Chemistry Chemical Physics 20, 9910–9921 (2018).

[3] J. Červený, J. Šalagovič, F. Muzika, D. Šafránek and I. Schreiber, Influence of Circadian Clocks on Optimal Regime of Central C-N Metabolism of Cyanobacteria. In: Cyanobacteria: From Basic Science to Applications, Edited by A.K. Mishra, D.N. Tiwari and A.N. Rai, Elsevier, London, 2019, pp. 193-206.

[4] J. Červený, M. A. Sinetova, L. Valledor, L. A. Sherman, L. Nedbal, Ultradian metabolic rhythm in the diazotrophic cyanobacterium Cyanothece sp. ATCC 51142. Proc. Natl. Acad. Sci. 110 (32), 13210–13215

(2013).

[5] J. Ross, I. Schreiber and M. O. Vlad, Determination of Complex Reaction Mechanisms, Oxford University Press, Inc., New York, 2006.

Copyright © American Institute of Chemical Engineers. All rights reserved.

576386 Metal Decorated Graphene Catalysts for the ORR: Effects of Nanoparticle Size, Composition, and Support Dopant on Activity

The development of next-generation materials and energy sources/fuels is driven by rising population, CO2 emissions, and standards of living. A promising technology to help alleviate these concerns is the development of the Proton Exchange Membrane Fuel Cell (PEMFC). Unfortunately, current state-of-the-art PEMFC technology requires catalyst materials that are platinum or platinum-group-metal (PGM) based at the electrode to facilitate reduction of oxygen via the Oxygen Reduction Reaction (ORR). To bring PEMFC technology to worldwide commercial development requires reducing the PGM loading (amount of PGM atoms in the catalyst) , as well as improve their efficiency. Recent advances in the literature have shown that formation of advanced Pt-based intermetallics and advanced structures (such as Kagome lattice materials) can both lower the Pt content and improve the overpotential and efficiency of the catalyst. In this work we present a competing approach- to use graphene as a doped/tuned support to modify the electronic structure and – hence- catalytic activity of small intermetallic transition metal nanoparticle (NP) catalysts of size: 4, 7, and 19 atoms.1-3 Our results to date show that the most promising combinations of size, composition, and dopant can fall near the predictive Volcano Plot derived for ORR activity on these systems.

The work described and presented was performed using modern, state-of the art plane-wave DFT calculations as implemented in the VASP 5.4x code deployed in the Materials Design MEDEA environment 2.21. Calculations and supercells for surface-mediate adsorption and catalytic phenomena involved dipole corrections parallel to the surface normal, vacuum spacing of at least 65% of the overall calculation cell height, and slab models of at least 6-8 Å or equivalence to 4 crystalline layers of the material(s) studied. Typical electronic convergence of the SCF was performed to at least 10-6 eV, and atomic force relaxations of the uppermost slab model atoms and adsorbed species was performed to less than 0.035 eV/ Å . Specific functional(s) used for the calculations in this study were the Van der Waals corrected opt-PBE.

Our results have indicated that ORR activity for 4,7, and 19 atom transition metal intermetallic nanoparticles on graphene and doped-graphene supports spans a *wide* range of descriptor values on the corresponding Volcano Plot. Careful selection of size, composition, and support-doping can cause predicted catalyst descriptor values to fall very close the to predicted volcano peak with similar or better overpotential than current state-of-the-art Pt-based ORR catalysts, while not using any PGM , or much less PGM atoms per proton-electron transfer. Our results form the basis of further ongoing study for stability and ability to be synthesized.

We will present results from DFT studies on the effects of composition, size, solvation, and support doping for the catalytic activity of Metal-Decorated-Graphene (MDG) nanocatalysts for the Oxygen Reduction Reaction (ORR). Conclusions and recommendations based on Volcano Plots are given, and counter-intuitive and non-linear results are discussed. To date our group is unique in publishing comprehensive studies on the possible activity of novel MDG catalysts for the ORR. This presentation summarizes our work in seeing the broad range to which the possible catalytic activity of these systems – if they can be synthesized- can yield for ORR electrochemistry; our results show equivalent or higher performance to state of the art Pt-based alloys and Kagome systems while theoretically using less Pt and other precious metals.

References

1. Wang, Z.; Ping, Y.; Fu, Q.; Pan, C., Preparation of Metal Nanoparticle Decorated Graphene Hybrid Composites: A Review. MRS Advances 2018, 3 (15-16), 849-854.

2. Lozano, T.; Rankin, R. B., Computational predictive design for metal-decorated-graphene size-specific subnanometer to nanometer ORR catalysts. Catalysis Today 2018, 312, 105-117.

3. Hu, M.; Yao, Z.; Wang, X., Graphene-Based Nanomaterials for Catalysis. Industrial & Engineering

Chemistry Research 2017, 56 (13), 3477-3502

4. Rankin, R.B, Lozano T; accepted/in print https://doi.org/10.3390/met9020227

5. Rankin, R.B. Lozano T; submitted April 2019.

Copyright © American Institute of Chemical Engineers. All rights reserved.

578226 Assembly of Gold Nanowires on Gold Nanopad Arrays: Simulation and Experiment

Nanowires (NWs) have a large aspect ratio and available surface area, which has made them beneficial for applications in biosensors, electronic/optic devices and power storage devices. However, realization of many of these applications requires NWs to be positioned and oriented through an assembly process. Hence, it is essential to develop assembly strategies and techniques to achieve desired structures. Self-assembly techniques can be utilized to assemble NWs over a wide range of sizes in the absence of an externally applied driving force (i.e., fluid flow or magnetic or electric fields). In this work, we explored NW assembly using a patterned NW-substrate interaction. Experimentally, silica-coated Au nanowires (diameter 340 nm, lengths of 2.4 μm and 4.4 μm) are observed to assemble into various different patterns on parallel aligned, Au nanopads (widths of 2.4, 4.5 and 4.8 μm), which create a series of "stripes" on a glass substrate. We develop a model of this system considering van der Waals (vdW) and electrostatic interactions among NW and vdW interactions between NW and the nanopads. Due to roughness, NW accumulate extra charge at their ends, and this plays an essential role in describing the assembly. Monte Carlo simulations of assembly were performed based on this model. The assembly patterns and properties for various NW agree well with experimental results. The simulations indicate that while the patterned vdW attraction generated by the Au nanopads plays a leading role in bringing NWs onto the pads, interparticle forces dictate NW ordering on the pads.

Copyright © American Institute of Chemical Engineers. All rights reserved.

569448 Role of Chloride in the Shape-Selective Growth of Ag Nanocrystals

Ag nanocrystals have promising applications in many cutting-edge technologies, which are highly dependent on their shapes. These nanomaterials are typically synthesized using the solution-phase polyol method, which involves the precursor $AgNO_3$ salt, the capping agent polyvinylpyrrolidone (PVP), and ethylene glycol as the solvent and reducing agent. Chloride and bromide are often added to achieve robust shape control. In the interest of achieving shape-selective syntheses, understanding the mechanistic origins of shape selectivity is important, but elusive.

Previous theoretical/experimental studies indicate that the preferential binding of PVP to Ag(100) can produce kinetic nanocubes of sufficiently large sizes. Small nanocubes can be synthesized experimentally when chloride is present, while theory predicts that truncated nanocubes are thermodynamically preferred in the presence of chloride and truncated octahedral are thermodynamically preferred without chloride. Here, we use dispersion-corrected density-functional theory to investigate the possibility that there is a synergy in the binding of PVP and chlorine to Ag surfaces, through which nanocubes emerge as the preferred thermodynamic shape.

We study the co-adsorption of PVP dimers and Cl to Ag(100) and Ag(111). We find that the adsorption of Cl influences the binding of PVP in a facet-selective way, such that Cl weakens the binding of PVP to Ag(111) at high Cl surface coverages, at which Cl continues strengthening the binding of PVP to Ag(100). We construct a contour plot of Ag-PVP-Cl surface energies as a function of the solution-phase Cl and PVP chemical potentials and find the associated Wulff shapes of the Ag nanocrystals. We see that truncated octahedra are thermodynamically preferred at low Cl concentrations / chemical potentials under full protection of PVP and Ag nanocubes are obtained at high Cl concentrations. Our results also indicate that Cl and PVP could work synergistically to produce five-fold twinned Ag nanowires under kinetic control.

Copyright © American Institute of Chemical Engineers. All rights reserved.

570781 Using Enhanced Sampling Methods to Study the Behavior of Osteocalcin on Mineral Surfaces

Biomineralization, the natural process by which organisms produce minerals, motivates the design of nature-inspired materials applicable in bone and dental regeneration.[1] Osteocalcin (dhOC), a natural non-collagenous protein abundantly found in bone and dentin, exhibits a high affinity for mineral surfaces such as calcium-phosphate rich hydroxyapatite (HAP) – a naturally occurring form of calcium apatite, and a major component of bone. It is understood the dhOC plays a prominent role in bone mineralization. However, the mechanisms and energetic driving forces that govern these interactions remain difficult to probe with experiments alone. It's affinity towards mineral surfaces has been prescribed to highly favorable binding between glutamic acid and calcium.[1] The presence of three potassium-dependent post-translational modification of glutamic acid (GLA) residues in the α-1 helix of dhOC are believed to be important for regulating and strengthening binding of dhOC to HAP[2]. Because modified, partially modified, and unmodified variations of dhOC are all naturally occurring, we are motivated to investigate how the structural or thermodynamic properties of dhOC α-1 change as a function of post-translational modifications. Further, the interactions between dhOC and its modifications on common implant materials such as silica (SiO_2) and titania (TiO_2) are much less understood, prompting additional interest in investigating these modifications as a function of surface effects.

This talk will highlight our application of enhanced sampling molecular dynamics methods to evaluate binding free energies and structural information concerning the α-1 helix of dhOC bound to HAP, TiO_2, and SiO_2 in unison with experimental studies. Using this comprehensive protocol, we hope to paint a complete picture of the of this protein's role in binding to organic materials.

References

1. Palmer, L. C., Newcomb, C. J., Kaltz, S. R., Spoerke, E. D. & Stupp, S. I. Biomimetic Systems for Hydroxyapatite Mineralization Inspired By Bone and Enamel. *Chem. Rev.* **108,** 4754–4783 (2008).
2. Malashkevich, V. N., Almo, S. C. & Dowd, T. L. X-ray Crystal Structure of Bovine 3 Glu-Osteocalcin. *Biochemistry* **52,** 8387–8392 (2013).

Copyright © American Institute of Chemical Engineers. All rights reserved.

566681 Adsorption and Disintegration of Surfactant Micelles at Metal-Water Interfaces: A Computational Investigation

Adsorption of surfactants at metal-water interfaces is useful for many technological applications, including corrosion inhibition and heterogeneous catalysis. Using atomistic simulations, we have studied how the adsorption behavior of surfactant molecules changes upon their micellization. We find that while un-aggregated surfactant molecules strongly adsorb onto metal surfaces, surfactant micelles experience a long-range repulsion from the surfaces. Upon overcoming the free energy barrier to adsorption, the adsorption free energy is related to the integrity of the micelles. The adsorption is weak when the micelles do not break, but the adsorption is strong when the micelles disintegrate in the adsorbed state.

Copyright © American Institute of Chemical Engineers. All rights reserved.

574631 A Molecular Dynamics Study to Investigate the Structural Features of Surfactant Aggregates Adsorbed at Mineral/Water Interfaces

Surfactants have long been employed to stabilize colloidal nanoparticles in suspensions through a combination of electrostatic and non-covalent interactions. Colloidal nanoparticles remain dispersed because of favorable interactions between surfactant assemblies such as micelles and bilayers, and the surface of these nanoparticles. Prior studies have discussed the structural and energetic features of these surfactant assemblies using experiments and molecular simulations. Despite recent advances, the formation mechanism for surfactant aggregates beginning with individual surfactant molecules at mineral/water interfaces (MWI) remains relatively unexplored. Herein, with the $CaCO_3$/water interface as a model system for the MWI, we use molecular dynamics simulations to rigorously investigate the mechanism by which surfactant aggregates like micelles and lamellar bilayers are formed. We also discuss the morphological features of these aggregates. We equilibrate our system through successive MD simulations in the NVT and NPT ensembles, following which we perform a production run in the NPT ensemble. Free energies from our MD simulations reveal the relative stabilities of micelles versus lamellar bilayers as a function of the surfactant concentration and environmental variables like temperature and counter-ion identity.

Copyright © American Institute of Chemical Engineers. All rights reserved.

562357 Fluid Transport in Shale Nanopores

Production of hydrocarbon in a hydraulically fractured shale starts from the desorption and diffusion of gas/oil from the matrix nanopores to the induced fractures and to the well. In shale nanopores, multiple components such as gas, oil, water, and super critical carbon dioxide ($scCO_2$) can coexist. Studying the interaction of these components with shale is critical to understanding the oil and gas transport in shale nanopores. We employed molecular simulations to investigate the properties of the fluid/shale system under reservoir conditions including gas adsorption, swelling, wettability, and transport. Our results indicate that kerogen preferentially retains CO_2 over CH_4, and that upon the adsorption of gas kerogen swells significantly. In addition, under ambient condition water partially wets the kerogen surface. However, in the $scCO_2$ atmosphere, water forms a spherical droplet at the kerogen interface. Our results also suggest that $scCO_2$ significantly enhances the flow of oil and gas in both organic and inorganic nanopores. Our work provides molecular-level insight into the production decline curve, chemo-mechanical properties of porous materials, and suggests a new mechanism and engineering approach to reduce energy consumption of a nanofluidic system.

Sandia National Laboratories is a multi-mission laboratory managed and operated by National Technology and Engineering Solutions of Sandia, LLC., a wholly owned subsidiary of Honeywell International, Inc., for the U.S. Department of Energy's National Nuclear Security Administration under contract DE-NA0003525. This research was funded by a DOE National Energy Technology Laboratory project.

Copyright © American Institute of Chemical Engineers. All rights reserved.

573751 Thermodynamic Techniques to Capture Non-Ideal Surfactant Assembly at Hard Nanoscale Interfaces

Surfactant behavior and structure is well studied with industrial applications. Adsorption isotherms of surfactants are critical in determining the relationship between the interfacial properties and structures of surfactants, but providing quantitative predictions of the isotherms remains challenging. This is especially true for adsorption at hard interfaces such as on 2D layered materials or on nanoparticles where simulation techniques developed for fluid-fluid interfaces do not apply. Here we predict non-ideal adsorption at a solid-solution interface with a molecular thermodynamic theory (MTT) model. Our approach uses MD simulations for the determination of MTT parameters from two thermodynamic integrations. Furthermore, the MTT/MD model provides atomistic insights into non-ideal behavior of surfactant molecules by capturing structural details of the surfactants at different concentrations. Our approach captures structural transitions from the ideal (non-interacting) case at low concentrations, monolayer assembly, and then to the critical surface adsorption concentration (CSAC, where hemi-micelles form) and finally through the CMC where micelles are stable in the bulk. We validate our model against the original MTT model by comparing predicted adsorption isotherms of a simplified surfactant system from both approaches. We further substantiate the applicability of our model in complex systems by providing adsorption isotherms of sodium dodecyl sulfate (SDS) in a water-graphene system, in agreement with experimental observations of the CSAC for the same system.

Copyright © American Institute of Chemical Engineers. All rights reserved.

565055 Impacts of Entropic Separation Effects

Adsorption processes using metal-organic frameworks – or nanoporous materials in general – are capable of replacing unsustainable separation processes like distillation or extraction to isolate hydrocarbons. Besides the well-known separation mechanisms based on enthalpic, kinetic or steric (i.e., sieving) differences, there is a further mechanism based on entropic differences. Those entropic effects describe the efficiency with which guest molecules can be arranged within the channels of nanoporous materials. In some cases, they lead to astonishing selectivities [1-3]. Unfortunately, these effects are seldom taken into account and seem to be virtually unknown.

The motivation behind this project is to contribute towards a better understanding of those entropic separation effects and indicating potentially interesting metal-organic frameworks, which are capable to efficiently solve a separation problem by means of entropic effects. To that end, grand-canonical Monte Carlo simulations were conducted to compute the adsorption isotherms of several hydrocarbon mixtures for a wide range of metal-organic frameworks and to visualise the arrangement of the investigated guest molecules within the channels of those materials.

For instance, in the event of separating hexane isomers – which is of great importance to the petrochemical industry – several metal-organic frameworks have been identified, whose selective adsorption of dibranched isomers is caused by entropic effects. Building upon these results, entropic effects seem to be capable of solving some of the major separation problems in energy and chemical engineering.

[1] Z. R. Herm et al., *Science*, **340**, p. 960 ff. (2013)
[2] A. Torres-Knoop, J. Heinen, R. Krishna, D. Dubbeldam, *Langmuir*, **31**, p. 3771 ff. (2015)
[3] R. Krishna, J. M. van Baten, *Sep. Purif. Technol.*, **55**, p. 246 ff. (2007)

Copyright © American Institute of Chemical Engineers. All rights reserved.

556746 Multiscale Electronic Coarse-Graining of Soft Materials Via Supervised Machine Learning

The computational design of soft materials with advanced electronic, ionic, or optical functionalities depends on the accurate treatment of (i) classical morphologies and (ii) quantum-mechanical electronic structure. This effort is challenging due to the multiscale nature of soft materials in which the strong coupling of disparate length and time scales mediates material performance. Traditionally, a hybrid classical and quantum-mechanical approach is employed, for which the sampling of conformational space can be demanding. Coarse-grained (CG) models can produce representative morphologies, but CG configurations must be back-mapped into atomistic representations to perform quantum-mechanical calculations, which is computationally intensive and inconsistent with the spatial resolution of the CG models.

Here, a supervised machine learning methodology, denoted as artificial neural network electronic coarse graining (ANN-ECG), is described in which the conformationally dependent electronic structure of a system is directly mapped to CG degrees of freedom. By averaging over decimated degrees of freedom, ANN-ECG accelerates simulations by orders of magnitude via eliminating backmapping and repeated quantum-chemical calculations. The approach is accurate, consistent with the CG spatial resolution, and can be used to derive physical intuition regarding collective variables that mediate electronically functional soft materials.

Copyright © American Institute of Chemical Engineers. All rights reserved.

575242 Molecular Simulations and Machine Learning for Multicomponent Adsorption

Industrial catalytic and separation processes use complex, multicomponent feedstocks, and often operate at high temperatures and pressures. Consequently, process design requires knowledge of multicomponent adsorption across "process space" – the multidimensional space of temperature, pressure, and composition. Measuring fluid properties and adsorption under *in situ* conditions is challenging – especially when attempting to cover wide composition ranges of multicomponent mixtures.

We performed high-throughput Monte Carlo simulations in the Gibbs ensemble to predict real fluid properties of liquid, vapor, and supercritical fluids in equilibrium with zeolite adsorbents. We examined unary adsorption and competitive adsorption of mixtures with up to eight components. These simulations provide numerous discrete data points that are used to fit an artificial neural network, which provides a single, self-consistent, continuous, and differentiable function describing the properties of all fluids and adsorbed phases across the full range of realistic process conditions. Properties and their derivatives from the neural network can enable accurate predictions for process optimization and control.

Copyright © American Institute of Chemical Engineers. All rights reserved.

567961 Designing the Phase Behavior and Microstructure of Nanocrystal Linker Gels

Gels made from networks of nanocrystals (NCs) linked by physical or chemical bonds are promising candidates for soft materials with tunable optical and rheological properties. A standard route to produce a gel is to rapidly cool a suspension of isotropically attractive particles, with gelation resulting from kinetic arrest during spinodal decomposition. However, such gels are inherently nonequilibrium, and the gel's properties inevitably change as it ages. Alternatively, "equilibrium" gels with open, homogeneous structures that are resilient to aging can be created by restricting the number of bonds that form between particles. Here, we report on equilibrium gelation controlled by the addition of a secondary "linker" macromolecule that mediates bonding between NCs. The phase diagrams of such mixtures were predicted using Wertheim's thermodynamic perturbation theory and compared to molecular dynamics simulations. Good agreement was obtained between the predictions and simulations, with the spinodal region depressed to lower NC densities by increased linker length at fixed linker-to-NC number ratio. The simulations further revealed that the linker length modulates the gel microstructure, which in turn controls its optical activity and rheology. We propose a strategy to simultaneously design the phase behavior and microstructure of the gel using blends of linkers with different lengths.

Copyright © American Institute of Chemical Engineers. All rights reserved.

577267 Multiscale Modeling and Enhanced Sampling to Examine Self-Assembly in Peptide and Peptoid Systems

Molecular self-assembly driven by non-covalent interactions is a central phenomenon that regulates a multitude of biological functions, from the formation of cell membranes, to the folding of polypeptide chains into proteins. Synthetically, the principles of molecular self-assembly are applied to produce crystals, polymers, and scaffolds for implants. To achieve greater control over manufacturing self-assembled materials, it is imperative to study driving forces that lead to the association of these molecules. As molecular assemblies occur over a gamut of length scales, it is useful to have models at different levels of complexity. In this talk, I will discuss distinct computational approaches that were implemented to elucidate the molecular assembly in two different systems.

First, I studied two biomineralizing peptides R5 and LKa14, which interact with orthosilicic acid to precipitate silica in solution. Both peptides have different secondary structures – R5 is random, whereas LKa14 is helical. However, they both produce silica that is morphologically similar. This suggests that there are dominant interpeptide interactions that cause peptides to assemble in a way that enables uniform templating of minerals. Using enhanced sampling methods to capture aggregation behavior occurring over long times, I compared the dominant driving forces leading to the assembly of these peptides. This knowledge can potentially help manufacture hierarchical nanomaterials in a predictive way, under ambient conditions.

Next, I systematically developed a coarse-grained model for peptoid systems using a bottom-up approach. Peptoids, positional isomers of peptides, are robust, synthetic, bio-inspired polymers that fold and assemble into a variety of nanostructures. An atomistic peptoid forcefield was developed recently that was shown to accurately reproduce local structures from quantum mechanical calculations. I used this forcefield to explicitly parameterize my CG model. To ensure reasonable transferability, the model was further parameterized using different combinations of hydrophobic and hydrophilic sidechain chemistries, chain lengths and temperatures. The mapping is similar to the MARTINI forcefield, where 4 heavy backbone atoms are mapped to a single CG bead. Self-assembled CG peptoids were compared with experimentally observed peptoid networks, to elucidate functional properties of the self-assembled aggregates arising from underlying molecular structure.

Copyright © American Institute of Chemical Engineers. All rights reserved.

574627 Modeling Thermodynamics and Kinetics at 2D Material Interfaces: Applications in Synthesis, Nanopore Formation, Wetting, and Catalysis

Two-dimensional (2D) materials, such as, graphene, transition metal dichalcogenides (TMDs) (*e.g.*, molybdenum disulfide (MoS_2)), hexagonal boron nitride (hBN), and layered metal oxyhydroxides (*e.g.*, nickel oxyhydroxide (NiOOH)) have recently received considerable attention, due to their exceptional optoelectronic, mechanical, barrier, and catalytic properties. However, a physical understanding of the controlled synthesis and interfacial behavior of 2D materials is still lacking. In this talk, I will discuss our research in several fundamental areas pertaining to modeling thermodynamics and kinetics at the interfaces formed by 2D materials:

1. **Mechanistic models for the chemical vapor deposition growth and etching of 2D materials:** I will discuss a generalized theory for the growth of TMD monolayers using chemical vapor deposition (CVD). The combined use of kinetic Monte Carlo (KMC) simulations and a transport model based on chemical engineering principles will be shown to enable the prediction of the experimentally observed shape and size evolution of the MoS_2 morphology inside a CVD reactor. Next, I will discuss the solution of the *Isomer Cataloging Problem (ICP)* for lattice nanopores in 2D materials. Combining density functional theory (DFT) calculations, KMC simulations, and chemical graph theory, I will present a catalog of unique, most-probable isomers of 2D lattice nanopores, demonstrating remarkable agreement with experimental transmission electron microscopy data for nanopores in graphene and hBN.
2. **Development of force field models to understand wetting and friction at 2D material surfaces:** I will outline the development of classical force fields for MoS_2 and hBN, for use in mechanical and interfacial applications, using DFT and lattice dynamics calculations. The force fields will be shown to predict the crystal structure and elastic constants of MoS_2 and hBN with good accuracy, and to correctly describe the mild hydrophilicity of the MoS_2 and hBN basal planes. I will also discuss the use of molecular dynamics (MD) simulations to understand the role of dispersion interactions, electrostatics, and entropy in the wetting and frictional properties of 2D materials. Further, I will discuss how our work led to some unexpected findings, including: (i) the dominance of dispersion interactions and entropy over electrostatics in determining the wettability and friction of 2D MoS_2, and (ii) that liquids with lower wettability can demonstrate higher friction on 2D hBN.
3. **Understanding catalysis at 2D material interfaces using first-principles simulations:** I will outline the use of conventional DFT with Hubbard-like corrections (DFT+U) and hybrid DFT with a fraction of exact exchange to understand the electrocatalytic behavior of pure NiOOH in terms of the oxygen evolution reaction (OER). By calculating Gibbs free energies for various elementary reactions occurring during the simple associative mechanism for water splitting, I will discuss the prediction of a lower bound for the overpotential exhibited by the surface for the OER. Finally, I will discuss the efficacy of dopants such as iron and manganese in enhancing the OER capability of NiOOH, including contrasting and rationalizing results from DFT+U and hybrid DFT.

In conclusion, the theoretical and simulation-based research work discussed in this talk should inform the controlled synthesis of 2D materials, and their use in various applications, including optoelectronic devices, mechanical composites, membranes for gas separation/water desalination, and electrocatalysis for splitting water molecules into oxygen and hydrogen.

Copyright © American Institute of Chemical Engineers. All rights reserved.

575885 Studying Inhomogeneous Ion-Containing Complex Fluids Using Polarizable Field-Theoretic Simulations

The growing interest in ion-containing polymeric materials with heterogeneous dielectric properties (salt-doped polymers, polyelectrolytes, ionomers, polymeric ionic liquids, etc.) has motivated the recent development of an equilibrium polarizable field theory, in which molecules are constructed out of beads that are granted polarizability via attached classical Drude oscillators. The incorporation of bead polarizabilities has the advantage that it self-consistently embeds dielectric response, van der Waals interactions and ion solvation energies into the theory, yet simulations of complex fluids based on such molecular models are rarely attempted. Here, by casting the statistical theory into field-theoretic form, it can be simulated without approximation using the complex Langevin field-theoretic simulation (CL-FTS) approach. This is crucial, since much of the emergent charge physics vanishes in the absence of electrostatic field fluctuations, rendering the mean-field approximation to the polarizable field theory inappropriate for most problems of interest. In this talk, I will summarize my recent work in applying the polarizable field theory to study i) the phase behavior of neat and salt-doped polymeric systems, and ii) the dielectric response of electrolyte solutions. This work reveals important aspects of the physics of ion-containing polarizable fluids which have not been previously anticipated in the literature.

Copyright © American Institute of Chemical Engineers. All rights reserved.

578395 The Role of Deformability in Determining the Structural and Mechanical Properties of Bubbles and Emulsions

We perform computational studies of jammed particle packings in two dimensions undergoing isotropic compression using the well-characterized soft particle (SP) model and the deformable particle (DP) model that we developed for compressed bubbles and emulsions. In the SP model, circular particles are allowed to overlap, generating purely repulsive forces. In the DP model, particles minimize their perimeter, while deforming at fixed area to avoid overlap during compression. We directly compare the structural and mechanical properties of jammed particle packings generated using the SP and DP models as a function of the true packing fraction ρ, instead of the reduced number density φ. We show that near jamming onset the excess contact number $\Delta z = z - z_J$ and shear modulus G scale as $\Delta\rho^{0.5}$ in the large system limit for both the SP and DP models, where $\Delta\rho = \rho - \rho_J$ and $z_J \approx 4$ and $\rho_J \approx 0.842$ are the values at jamming onset. Δz and G for the SP and DP models begin to differ for $\rho \approx 0.88$. In this regime, $\Delta z \sim G$ can be described by a sum of two power-laws in $\Delta\rho$, i.e. $\Delta z \sim G \sim C_0\Delta\rho^{0.5} + C_1\Delta\rho^{1.0}$ to lowest order. We show that the ratio C_1/C_0 is much larger for the DP model compared to to that for the SP model. We also characterize the void space in jammed packings as a function of ρ. We find that, unlike the SP model, the DP model is able to describe the formation of Plateau borders as the system approaches $\rho = 1$. We further show that the results for z and the shape factor A versus ρ for the DP model agree with recent experimental studies of compressed foams and emulsions.

Copyright © American Institute of Chemical Engineers. All rights reserved.

568091 Unraveling the Quantum Mechanical Catalytic Action of Methyltransferases with GPU-Accelerated Large-Scale Electronic Structure

Methyltransferases play key roles in biological processes, including epigenetic modifications, protein repair, natural product biosynthesis, and neurotransmitter regulation. Although methyl transfer is believed to occur with an S_N2 mechanism for S-adenosyl-L-methionine (SAM)-dependent class I methyltransfases, the role of the greater enzyme environment remains under debated. Electrostatics, charge transfer, dynamics, and C-H···O hydrogen bonding have all been suggested to contribute to reactivity. Here, we employ GPU-accelerated electronic structure calculations to explore the quantum mechanical catalytic origin of methyltransferase. We performed large-scale QM/MM calculations (~500-1000 QM atoms) on four representative class I methyltransferases, including 2'-O-ribose (CMTr1), phosphoethanolamine (PfPMT), L-isoaspartyl (PIMT), and C3-6-carboxymethyl-5-methyl-4-hydroxy-2-pyridinol (HcgC) methyltransferases. We observe that favorable electrostatic interaction between the methyl donor SAM and the substrate is largely maintained from reactant to transition state region, which is resulted from charge transfer between the reacting molecule (SAM + substrate) and their greater protein environment. CHO(N) hydrogen bonds, albeit weak, were observed in all methyltransferases studied here. From reactant to transition state region, the hydrogen bonding interactions were found to be slightly enhanced in CMTr1, PfPMT, and HcgC, but diminished in PIMT. Our work demonstrates the promising utilization of GPU-accelerated quantum mechanical simulation to understand fundamental biochemical problems.

Copyright © American Institute of Chemical Engineers. All rights reserved.

572058 Molecular Simulations of Biomimetic Assembly

Clarifying the driving mechanisms involved in various biological processes (i.e., protein aggregation, membrane fusion, etc.) has become one of the main goals in our scientific community. One of the promising directions for this task is by performing molecular simulations of analogous biomimetic assembly. In this presentation, I correspondingly discuss the two main research routes which I am pursuing: Particularly, together with my colleagues, I investigate hydration forces between colloids and phase transitions in polymers. First, in the colloidal project, we foremost resolve a paradox that has existed for a half of a century between numerical predictions and empirical measurements regarding the hydrophobic force: We specifically find a continuous transition between the association of small hydrophobes (below ~1 nm in size), which assemble entropically via a weak oscillatory force, and the association of large hydrophobes (above ~1 nm in size), which assemble enthalpically via a strong monotonic force; we explain this length-scale crossover in terms of the ability of a water solvent in forming a tetrahedral network around nonpolar solutes [1-2]. Besides, we also systematically transform these biomimetic colloids into hydrophiles, finding a multifaceted energy-scale crossover for the hydrophilic force [3]. Second, in the polymeric project, we examine a bead-spring chain that embodies a coil-crystal transition: Specifically via perturbation theory, we show that such a biomimetic polymer characterizes the universal kinetics associated with the force spectroscopy of various biological chains [4]. We continue by studying the influence of this coil-crystal transition of a single polymer on the corresponding aggregation propensity of many polymers. On a practical level, our molecular simulations for the biomimetic assembly of colloids and polymers can be of much aid in the design of pharmaceuticals.

[1] A. Chaimovich and M.S. Shell (PRE, 2013).
[2] A. Chaimovich and M.S. Shell (PRE, 2014).
[3] A. Chaimovich, R.R. Lele, M.S. Shell, J. Israelachvili, and M. Valtiner (In preparation).
[4] A. Chaimovich, C. Leitold, and C. Dellago (In preparation).

Copyright © American Institute of Chemical Engineers. All rights reserved.

576085 A Versatile Coarse-Grained Approach for Simulating Multi-Component Polymers: Field-Accelerated Monte Carlo Simulation

Coarse-grained simulations have played a crucial role in advancing our understanding of polymer behaviors. The success owes to one unique feature of polymers -- their generic and universal behaviors are to a large extent insensitive to interaction details at the atomistic level. The reduced computational cost enables coarse-grained simulations to access the experimentally relevant degree of polymerization and structural length scale. Coarse-grained simulations can be generally categorized into the particle-based and field-based approach. The particle-based method provides more microscopic details and higher spatial resolution for a given model, while the field-based approach enjoys better computational efficiency in simulating dense and inhomogeneous systems. Here we present a hybrid particle-field scheme -- Field-Accelerated Monte Carlo (FAMC) Simulation, that aims to take advantage of both approaches. The FAMC scheme offers the flexibility of computing interactions in the simulated system in the particle- or field-based fashion depending on their specific types. The choice between the two representations can be made based on physical considerations of the strength and the magnitude of fluctuation of the interaction energy. Applications of FAMC to two polymer systems are discussed to illustrate the versatility of the method. The first example is a microphase separating diblock copolymer melt containing monomers that can form hydrogen bonds. A small fraction of these hydrogen bonding "moieties" could induce drastic changes in the free energy landscape, leading to abruptly different structural and dynamical behaviors. This requires simulations to simultaneously capture the microscopics of the reversible bonding (difficult to achieve in a full field-based approach) in the presence of inhomogeneous microstructures (computationally expensive for a full particle-based approach). Another application is to simulate dilute polymer solutions with explicit solvents. Such simulations are computationally demanding for a full particle-based method due to the need for computing a large number of solvent-solvent interactions. A mean field approach, on the other hand, is not sufficient to capture the large magnitude fluctuations as a result of small polymer concentration. We show that in simulating both systems, the FAMC method offers the optimal combination of accuracy and computational efficiency. A prospective outlook on the FAMC's applications in polymer research will also be discussed at the end of the presentation.

Copyright © American Institute of Chemical Engineers. All rights reserved.

577216 Parameterization of Coarse-Grained Models for Block Copolymers: Approaches to Simulating Real Polymer Systems

Block copolymers have received significant attention due to their ability to naturally self-assemble into a variety of complex nano- and micro-structured phases. The ability to form structure via thermodynamic driving forces rather than by requiring external manipulation has made them an interesting and potentially critically important material in a variety of applications including organic photovoltaics, semiconductors, optical coatings, and catalysts. In understanding and designing such block copolymer systems, polymer simulation techniques including self-consistent field theory (SCFT), Monte Carlo, and molecular dynamics methods have proven extremely useful. For example, coarse-grained models of block copolymers and the simulation of their microphase separation have provided valuable insight into the design and understanding of materials and processes for directed self-assembly (DSA) in the semiconductor lithography field. However, the majority of polymer models used in these simulation studies are relatively simple, which limits their utility and ability to represent real polymer systems. For example, many such studies have used the simple Flory-Huggins interaction parameter (χ) to describe the inter-block interactions. This approach fails to account for realistic polymer complexities such as asymmetry in block cohesive energy density. As a result, in many cases no attempt is even made to connect the modeling work to real polymer systems.

As the applications for block copolymers continue to progress and the need arises to utilize modeling and simulation methods to go beyond a qualitative understanding of phenomena to the design and simulation of real polymer systems, there is a critical need to develop models and methods for parameterizing them that can reproduce the behavior of real polymer systems. For example, while much has been learned about the DSA of block copolymers like poly(styrene-b-methyl methacrylate) from simulation studies utilizing the simplistic Flory-Huggins interaction model, the field is moving to higher χ polymers which possess differences in physiochemical properties such as block density and block cohesive energy density; unfortunately, such simple models fail to reproduce the proper behavior. While atomistically detailed molecular dynamics simulations provide an approach to more faithfully represent the important physiochemical behavior that would allow for the proper modeling of such systems, the general simulation size required to represent microphase-separated systems such as these block copolymers prohibits this approach in a practical sense due to the speed and memory limitations of current computational systems. Coarse-grained analogs instead offer a solution to permit simulation at the large length and time scales required, but this technique also requires the development of such coarse-grained models that can faithfully represent the important chemistry and physics of the real polymer systems of interest. In general, there is still no single widely accepted method for parameterizing such course-grained block copolymer models to reflect the atomic details and physical property behavior of a given specific polymer. As a result, in this work a systematic procedure to parameterize course-grained molecular dynamics models has been developed that utilizes data from a combination of experimental property measurements and detailed atomistic simulations. The resulting coarse-grained models include energetic parameters that go beyond the overly simplistic Flory-Huggins χ parameter to allow for the representation of polymers whose behavior is complicated by effects such as disparities in the cohesive energy density of the copolymer constituent blocks. This energetic interaction asymmetry between blocks is common among high-χ BCPs, and in particular it can explain some of the unique process behaviors such as the rotation of lamellae to be parallel to an interface and the resulting need for topcoat layers to prevent such rotation in DSA processes being explored for semiconductor device fabrication.

In this work, the methodology developed for coarse-graining and model parameterization has been applied to producing a coarse-grained model for the relatively high-χ poly(styrene-b-isoprene) system due to both: (1) the extensive body of available data on its phase behavior and other physiochemical properties and (2) the aforementioned block density and cohesive energy density differences in the PS-b-PI system. The ability of the coarse-grained model to reproduce important results such as the phase diagram for this system will be shown as one validation of this approach.

Copyright © American Institute of Chemical Engineers. All rights reserved.

577384 Application of the Energy Renormalization Method to a Coarse-Grained Polymer Model with a Dissipative Potential

Advanced manufacturing demands faster and more accurate molecular models for the development of new polymeric materials for applications such as lightweight vehicle components. Atomistic-level information is required to predict chemistry-specific temperature dependence such as the transition from a melt to a glass, but fully atomistic simulation systems are computationally expensive. Numerous coarse-graining (CG) simulation approaches have been developed to study polymers by lumping several atoms into fewer CG sites and enable the study of polymer dynamics over reasonable timescales. However, most current CG models are limited to a specific state point (i.e., they are not temperature-transferable), rendering them unable to predict properties during manufacturing or operating conditions. Here, we build on a recently developed temperature-transferable, coarse-graining algorithm that captures chemistry-specific dynamics from the melt to the glass, the Energy Renormalization (ER) scheme [1,2]. We study the extension of the ER scheme to higher levels of CG by additionally incorporating a dissipative term typically found in higher-level CG models into the molecular potential. We demonstrate that such a higher-level CG scheme still retains temperature transferability (i.e., recovery of temperature-dependent polymer dynamics) and enables improved recovery of structure.

References

[1] Wenjie Xia, Jake Song, Cheol Jeong, David D. Hsu, Frederick R. Phelan Jr., Jack F. Douglas, Sinan Keten, “Energy-Renormalization for Achieving Temperature Transferable Coarse-Graining of Polymer Dynamics,” Macromolecules, 50 (21), pp 8787–8796, (2017).

[2] Wenjie Xia, Jake Song, Nitin H. Krishnamurthy, Frederick R. Phelan Jr., Sinan Keten, Jack F. Douglas, “Energy Renormalization to Coarse-Graining of the Dynamics of a Model Glass-Forming Liquid,” Journal of Physical Chemistry B, 122 (6), pp 2040–2045 (2018).

Copyright © American Institute of Chemical Engineers. All rights reserved.

574889 A Systematic Multiscale Coarse-Grained Model of Actin Filament Networks

Actin cytoskeletal network, the fundamental framework of a cell, is critical for a cell's ability to move, divide, and transport cargo. Actin filaments constantly assemble, disassemble, sever, bundle and form branches in the presence of a number of actin-binding proteins (ABPs) giving rise to a dynamic network. The ABP mediated remodeling of these networks is governed by conformation-dependent interactions between actin and the ABPs. These conformational changes typically involve multiple timescales and therefore, simulating the entire network at a cellular level in atomistic resolution is computationally infeasible. In this regard, a multiscale modeling framework was previously developed in our group *[Katkar et al., Biophysical Journal, 115, 1589-1602, 2018]* to gain insights into nucleotide hydrolysis in a single actin filament.

In this work, we extend the framework to systematically simulate the dynamic remodeling of actin filament networks, both in the absence and presence of ABPs such as the arp23 complex. We adopt the Ultra-Coarse-Grained (UCG) methodology where the coarse-grained (CG) models undergo stochastic transitions between different states based on CG particle conformations during the simulations. Here, the transition between states is used to mimic processes that actin subunits undergo, such as polymerization, depolymerization, nucleotide hydrolysis, binding to ABPs etc. The model is parameterized using short atomistic simulations corresponding to each state and the average transition rates are tuned to agree with known macroscopic experimental rates. Using this model, we investigate the filament assembly under varying conditions of actin and ABP concentrations. Our systematic parameterization strategy allows us to make predictions pertaining to the cooperative effects between nucleotide hydrolysis and filament network remodeling.

Copyright © American Institute of Chemical Engineers. All rights reserved.

565452 Modeling Inter-Colloidal Interactions Induced By Adsorption of Mobile Telechelic Polymers Onto Particle Surfaces

Telechelic polymers can physically absorb onto colloidal surfaces, forming loops on single particles and loops between pairs of particles, which in turn induce effective interactions between colloids. We combine self-consistent field theory and a modified Derjaguin approximation to predict the polymer-induced colloidal interactions and the non-uniform distributions of loops and bridges of telechelic polymers adsorbed onto particle surfaces when they are compressed or stretched as a function of inter-particle distance. We validate our approach by comparing its predictions to those of Brownian dynamics simulations. We also determine the dependence of inter-colloidal interactions on particle size and surface coverage as well as molecular structures of telechelic polymers, including chain length and missing associating ends, which are important parameters in the design of commercial latex coatings. By mapping the predicted inter-particle interaction strengths to Baxter temperatures, we can quantitatively predict the phase behaviors of mixtures of colloids and telechelic polymers.

Copyright © American Institute of Chemical Engineers. All rights reserved.

569479 Performance Mechanisms of Viscosity Modifiers Using Coarse-Grained Molecular Dynamics

Engine oils are constantly exposed to multiple drive cycles as well as diverse ambient conditions during non-operative periods, but their viscometric performance is expected to be uniform and robust across different conditions. For high entrainment speeds, temperatures, and pressures, viscosity must be high enough to contribute to wear prevention within elastohydrodynamic and mixed tribological regimes. However, the oil also must be thin enough to be pumpable at low temperatures to assist cold starting and have a high viscosity index to promote fuel-economy. This effect is commonly achieved through the introduction of viscosity modifiers (VMs), which are high molecular weight (MW) polymers. The traditional belief is that the performance mechanism of VMs follows the coil expansion model (Selby, *ALSE Trans.*, **1**, 68-81, 1958). As temperature increases, the solvency power of the oil increases causing the polymers to swell or uncoil, increasing viscosity. However, experimental evidence has suggested that this view may be inadequate for all VMs. While poly(alkyl methacrylate)s (PAMAs) show monotonic viscosity increases with temperature, other VMs (*e.g.*, olefin copolymers (OCPs)) do not always show the same trend (Müller, *Tribol. Int.*, **11**, 189, 1978 and Covitch and Trickett, *Adv. Chem. Eng. Sci.*, **5**, 134, 2015).

Due to the macromolecular size of polymers, fully-atomistic simulations of VMs are too computationally intensive to study in dilute solution for MWs greater than about 2,000. We show that systematic coarse-graining allows us to accurately calculate structural properties of VMs with MWs used in engine oils (greater than 100,000). Using iterative Boltzmann inversion (Reith *et al.*, *J Comput. Chem.*, 1624, **24**, 2003), we develop coarse-grained force-fields for OCPs and PAMAs, two commonly used VMs in the lubrication industry. Using coarse-grained molecular dynamics, we calculate the polymer's radius of gyration and hydrodynamic radius in different organic solvents.

Copyright © American Institute of Chemical Engineers. All rights reserved.

575633 The Effect of Nucleoid Associated Proteins on DNA Supercoiling

DNA compaction is critical in order for a long DNA strand to fit into a much smaller cell. This can be done in a number of ways including looping, compartment segregation, and supercoiling. The latter occurs when DNA is torsionally stressed and forms large-scale writhed structures by wrapping around itself to relieve this stress. In prokaryotic cells, nucleoid associated proteins (NAPs) play an important role in DNA organization and compaction by manipulating the shape and structure of the DNA. These NAPs act by bending or twisting DNA at the local scale, and we hypothesize that these local effects strongly impact the stability and structure of DNA supercoils.

We use a combination of coarse-grained simulation of NAPs and DNA and theory to investigate the effect that these proteins have on DNA supercoiling behavior. These proteins are preferentially bound to a bent DNA site via the binding and unbinding energy landscape. Additionally, these proteins change the local equilibrium angle upon binding, creating localized kinks in the DNA strand. We are able to capture experimental observations with our coarse-grained model. We can take this information to inform a theory that accounts for the effect that proteins have on the supercoiled system, such as the extension behavior, bending energy, and excluded volume that are dependent on concentration and force.

Copyright © American Institute of Chemical Engineers. All rights reserved.

568227 Unraveling Structural Motifs of Heterogeneous Chromatin through Coarse-Grained Modeling

Chromatin is a unique biopolymer consisting of repeats of nucleosomes connected by flanking segments of short linker-DNA. In previous work, we developed a bottom-up Brownian dynamics modeling approach to unveil the mesoscopic structure of chromatin. Through this model, we are able to identify the delicate interplay between DNA flexibility and the anisotropic inter-nucleosome interactions, and how each plays a role on the structure of chromatin. In the ideal case (homogeneous linker-length), chromatin is comprised of ordered trinucleosome motifs. However, in the cell nucleus nucleosomes diffuse along the DNA, generating heterogeneous fibers. Here, we assess chromatin structure subject to nucleosome dynamics through structural and thermodynamic analysis. We first demonstrate that the structure of heterogeneous fibers of various linker-DNA distributions is disordered, yet occupies the same volume. We then utilize free energy methods to determine the relative thermodynamic stability between homogeneous and heterogeneous chromatin.

Copyright © American Institute of Chemical Engineers. All rights reserved.

558909 Applications of the Telegraph Model for Nanochannel Confined DNA

The study of nanochannel-confined polymers, in particular DNA, has received substantial attention in the past decade owing to its application in the field of genome mapping. Recently, a weakly-correlated telegraph model was proposed to describe the extension of the DNA molecules stretched under confinement using a single scaling parameter, and this model was solved asymptotically in the limit relevant to genome mapping. We will present an analysis of the applicability of the telegraph model to both simulation and experimental data. On the computational front, we have compared the predicted distribution of fractional extensions obtained by an asymptotic solution to the telegraph model to computational results using pruned-enriched Rosenbluth method (PERM) simulations for the same conditions. We find good agreement between the two methods. On the experimental front, we used recent results of a large dataset of fractional extension values as a function of the ionic strength to study the telegraph model predictions spanning several decades in its scaling parameter. We find good agreement at the smaller values of the scaling parameter where the telegraph model is expected to hold. The agreement between the model and experiments is considerably worse at larger values of the scaling parameter, which correspond to small effective channel sizes.

Copyright © American Institute of Chemical Engineers. All rights reserved.

576630 Probing the Sensitivity of Point Charge Potentials in an Attempt to Improve Their Ability to Rank the Stability of Pharmaceutical Polymorphs

Temperature and entropic effects cannot be neglected when ranking the stability of organic polymorphs. Enormous effort has been put into structure prediction to seek out new crystalline phases for organic materials, with much of the stability ranking being an afterthought. Structure predictions have relied on lattice energy to determine the crystalline stability despite numerous experimental and computational studies that have shown that enantiotropic behavior is prevalent. In particular, one study has shown that 20% of a set of > 500 experimentally know polymorphs re-rank with temperature.

The bottleneck preventing free energy stability rankings in CSPs is largely due to the availability of accurate and efficient force fields. We have developed and tested molecular and lattice dynamic approaches to include temperature and entropic effects of enantiotropic polymorphs. For both approaches, we used point-charge potentials to compute the free energy of 12 systems. We found that the entropy differences using the point charge potential was the correct sign for 100% of the systems relative to experiment, but the enthalpic stability between polymorphs was only the correct sign for 58% of the systems. While more complex hamiltonians could be used to correct these issues the increase in computational expense becomes disfavorable, especially if added to a CSP.

We hypothesize that point charge potentials can be tuned with simple scaling factors improve the lattice energy, without negative implications on the entropic contributions. Initial results have shown that the entropic contributions to free energy stability are less sensitive to the potential used than the lattice energy. To better understand the tunability of point charge potentials we will look at how scaling the intermolecular interactions independently affect the enthalpy and entropy differences of the crystalline systems previously studied. The end goal would be to take the lattice energy rankings of a CSP performed with a highly level of electronic theory, tune the intermolecular interaction to minimize the error in the lattice energy stability, and then consider the effects of temperature and entropy. I propose to present the results tested to prove or disprove that hypothesis.

Copyright © American Institute of Chemical Engineers. All rights reserved.

552602 Crystal Structure Prediction Applications - Going Beyond a Stable Form Selection

Crystal structure prediction (CSP) in pharmaceutical industry is typically performed to support commercial solid form selection of a freebase API. However, the variety of tasks of solid state chemistry in industry is much broader than that important application. The presentation will cover some virtual screening CSP-based applications such as poor solubility improvement during lead optimization and coformer screening for cocrystallization.

Copyright © American Institute of Chemical Engineers. All rights reserved.

563344 Accurate and Efficient Lattice Energy Models for Industry-Oriented Crystal Structure Prediction

Many organic molecules of industrial interest exhibit crystalline polymorphism, forming multiple solid-state structures, each displaying different physico-chemical properties. Consequently, understanding the crystal energy landscape of a commercial drug or chemical can be of great importance, especially if the compound is delivered in solid form, as is the case with suspended agrochemicals or oral dosage forms for pharmaceuticals. The discovery of new polymorphs is thus desirable from an R&D perspective and can help to avoid potentially disastrous situations from an operational standpoint. These observations have motivated the development of methods for polymorph prediction that can now be applied to molecules of industrial relevance, provided that conformational flexibility is limited to a few degrees of freedom.[1].

Recent blind tests [2,3] organised by the CCDC have highlighted the growing capabilities of crystal structure prediction (CSP) for increasingly complex systems, but what is often considered a success in a blind test does not always provide sufficient information to complement experimental solid-form screening efforts in an effective way. In addition to these concerns, ensuring that CSP can be used in the early stages of product development requires that the crystal energy landscape can be mapped quickly and efficiently [4]. Despite the requirement for efficient methods, the use of computationally expensive periodic density functional theory (DFT) calculations has become a staple for many practitioners of CSP. Although this typically achieves greater accuracy, the computational cost may often be prohibitively large for the study of systems of industrial relevance in the early stages of product development.

In this talk, we investigate the role of parameter estimation and adaptive force fields in the development of bespoke *ab initio* potentials. Models developed in this way can approach the accuracy of periodic DFT calculations with a computational cost reduction typically around 6 orders of magnitude. This is partly achieved a) through the rigorous construction of training sets from theoretical calculations and b) by taking cues from successful developments in DFT dispersion corrections [5,6] over recent years to improve the description of intermolecular interactions. Overall, significant improvements in the accuracy of the predictions of the energy and geometry of organic solids can be obtained, over conventional force fields, increasing both the speed and reliability of CSP as applied in industrial applications.

[1] Nyman, Jonas, and Susan M. Reutzel-Edens. "Crystal structure prediction is changing from basic science to applied technology." *Faraday discussions* 211 (2018): 459-476.

[2] Bardwell, David A., et al. "Towards crystal structure prediction of complex organic compounds–a report on the fifth blind test." *Acta Crystallographica Section B: Structural Science*67.6 (2011): 535-551.

[3] Reilly, Anthony M., et al. "Report on the sixth blind test of organic crystal structure prediction methods." *Acta Crystallographica Section B: Structural Science, Crystal Engineering and Materials* 72.4 (2016): 439-459.

[4] Price, Sarah L., and Susan M. Reutzel-Edens. "The potential of computed crystal energy landscapes to aid solid-form development." *Drug Discovery Today* 21.6 (2016): 912-923.

[5] Grimme, Stefan, et al. "A consistent and accurate ab initio parametrization of density functional dispersion correction (DFT-D) for the 94 elements H-Pu." *The Journal of chemical physics* 132.15 (2010): 154104.

[6] Tkatchenko, Alexandre, and Matthias Scheffler. "Accurate molecular van der Waals interactions from ground-state electron density and free-atom reference data." *Physical review letters* 102.7 (2009): 073005.

Copyright © American Institute of Chemical Engineers. All rights reserved.

567777 How Many Ritonavir Cases Are Still out There?

On rare occasions, a crystal structure that has been prepared for years is unexpectedly "superseded" by a thermodynamically more stable polymorph, rendering the initial crystal structure almost impossible to obtain—a so-called "disappearing polymorph" [1]. Famous cases include the drug compounds Ritonavir and Rotigotine: The thermodynamically more stable polymorph turned out to be substantially less soluble and required a costly reformulation before a product with comparable specifications could be manufactured again.

On a case-by-case basis, the probability of a kinetically hindered but thermodynamically stable polymorph having been missed in an experimental polymorph screen can be assessed by a computational crystal structure prediction study: the calculations do not suffer from kinetics and will in principle enumerate all possible crystal structures for a given compound; comparison against the experimental structures identifies any missed polymorphs. In a recent paper, we presented the results of 41 such computational crystal structure prediction studies to draw up statistics regarding the number of hidden Ritonavir cases [2]. The main finding of the paper was that between 15 and 45% of all experimental crystal structures are in fact thermodynamically unstable.

The large spread in our estimate was due to the computational error in the calculation of the relative free energies of the predicted polymorphs. Even though the computational error is estimated to be as small as 0.5 kcal/mol, about half the size of the gold standard referred to as "chemical accuracy", the experimental energy differences between polymorphs are also around the 0.5 kcal/mol mark, and to reduce the large spread in our estimate an even more accurate energy model was needed.

The shortcomings of our old energy model such as neglect of temperature, relatively poor accuracy of the exchange part of the functional used in the DFT calculations, and the assumption that Van der Waals interactions are pairwise additive are well documented, and in this contribution we will present the results of an energy model in which these shortcomings have been addressed [3]. With the availability of more accurate relative lattice energies, we are able to narrow down our estimate of how many experimental crystal structures represent a thermodynamically unstable polymorph at ambient conditions.

References

[1] D.-K. Bučar, R. W. Lancaster & J. Bernstein (2015), "Disappearing Polymorphs Revisited" *Angew. Chem., Int. Ed.* **54**, 6972-6993.

[2] M. A. Neumann & J. van de Streek (2018), "How Many Ritonavir Cases Are There Still out There?" *Faraday Discuss.* **211**, 441-458.

[3] J. Hoja, Hsin-Yu Ko, M. A. Neumann, R. Car, R. A. DiStasio Jr. & A. Tkatchenko (2019), "Reliable and Practical Computational Description of Molecular Crystal Polymorphs" *Sci. Adv.* **5**, eaau3338.

Copyright © American Institute of Chemical Engineers. All rights reserved.

562227 A Comparison of Methods for Capturing Hydrate-Anhydrate Transition Thermodynamics

A key milestone in the small molecule drug development pipeline is the selection of a solid form for both research and commercial development. One of the main deliverables from the solid form selection process is to identify a solid form of the Active Pharmaceutical Ingredient (API) that is physically stable and unlikely to undergo a solid form change during a later stage of manufacturing, transportation, and storage. One significant physical transformation that can occur in a drug product is a conversion from an anhydrous to a hydrate form because a hydrate can have a lower aqueous solubility and dissolution rate by over an order of magnitude relative to the corresponding anhydrous form.

Computational models of hydrate-anhydrate transitions can reduce the risk of unintended transformations during late-stage development. Recent molecular modeling approaches have been developed to compute the stability of anhydrous polymorphs pairs over a range of experimentally relevant temperatures. Here we extend this work to the comparison of anhydrous forms to their corresponding hydrates over a range of temperature as well as ambient humidity levels. The stabilities of multiple different small molecule hydrate-anhydrate pairs are computed using a variety of classical potentials including OPLS, GAFF, and the polarizable AMOEBA force-field. Additionally, the interstitial waters are modeled with the TIP3P, TIP4P, and AMOEBA water models. Overall, we find that the estimated entropy differences between forms are significantly less sensitive to the force-field than the corresponding enthalpy differences. In addition, the entropy differences are in decent agreement with experimental measurements derived from coexistence points. This result suggests that quantitative prediction of the transition temperature could be achieved in conjunction with a sufficiently accurate enthalpy difference estimate at a single temperature.

Copyright © American Institute of Chemical Engineers. All rights reserved.

583064 Genarris 2.0: A Random Structure Generator for Molecular Crystals

Genarris 2.0 is an open source Python package for performing configuration space screening by random structure generation for crystals of molecules with no bond-rotational degrees of freedom. The target unit cell volume is determined by estimating the molecular volume. Crystal structures are then generated in all space groups compatible with the requested number of molecules per cell (Z) with one molecule in the asymmetric unit (Z'=1), including those with special Wyckoff positions. To avoid generating unphysical structures, constraints are applied on the intermolecular distances, based on van der Waals radii. Special settings are applied for strong hydrogen bonds, which are automatically detected. Once an initial dataset of several thousand structures is generated, a smaller curated dataset may be selected based on user-defined quality and diversity criteria to serve as an initial guess for structure search algorithms or as a training set for machine learning. For clustering based on packing motif similarity Genarris uses the affinity propagation machine learning algorithm with either a relative coordinate descriptor (RCD) or a radial distribution function (RDF) representation.

Copyright © American Institute of Chemical Engineers. All rights reserved.

583911 Crystal Structure Prediction in Lead Optimization Phase of Drug Discovery

Crystal structure predictions (CSP) has become a very valuable tool to explore crystal interactions in early discovery phase. At AstraZeneca we have developed a fast and accurate protocol that allows for applying CSP already in the lead optimization phase. The protocol will be described and compared to the most accurate protocols published. A few case studies will be presented to illustrate the speed and accuracy of the protocol.

Copyright © American Institute of Chemical Engineers. All rights reserved.

579605 Data Science for Assembly Engineering

Discovery and design of new materials able to self assemble from nanoscale building blocks are becoming increasingly enabled by large-scale molecular simulation. Aided by fast simulation codes leveraging powerful computer architectures, an unprecedented amount of data can be generated in the blink of an eye, shifting the effort and focus of the computational scientist from the simulation to the data. How do we manage so much data, and what do we do with it when we have it? In this talk, we discuss the applications of data science and data-driven thinking to molecular and materials simulation. Although we do so in the context of assembly engineering of soft matter, the tools and techniques discussed are general and applicable to a wide range of problems. We present applications of machine learning to automated, structure identification of complex colloidal crystals, high-throughput mapping of phase diagrams, the study of kinetic pathways between fluid and solid phases, and the discovery of previously elusive design rules and structure-property relationships.

We also discuss new tools we've developed to support these studies. We present *pythia*, a wrapper library for use with data analysis tools that generates descriptors of particle systems for ML. We present *signac*, an open-source data management framework that facilitates the management of large and heterogeneous data spaces by creating a well-defined, indexable storage layout for data and metadata. *signac* is lightweight and manages computational workflows on platforms ranging from laptops to supercomputers. *signac* supports HDF5 integration for rapid access to large numerical data arrays, has new tools to import and export data spaces for long-term archival and publication of data sets; and integrates seamlessly with the scientific Python ecosystem for use with Jupyter notebooks and more. *signac*is application agnostic and has been used for molecular simulations, quantum chemistry, photonics, computational fluid dynamics, machine learning, graph mining, and even organizing experimental data.

Contributors include Carl Simon Adorf, Joshua A. Anderson, Chengyu Dai, Bradley Dice, Yina Geng, Vyas Ramasubramani, and Matthew Spellings.

Copyright © American Institute of Chemical Engineers. All rights reserved.

580104 A Computational Roadmap for Better Electronic Noses

Most gas sensors can only detect a single gas species (e.g., a household carbon monoxide detector), with a small subset able to distinguish among a few gas species (e.g., commercial "electronic nose" sensors). However, few attempts have been made – and none successfully – at creating a device that could detect a very broad spectrum of gases under a wide range of operating conditions. This seemingly challenging task is carried out every day by our own biological noses, however. Creating an artificial nose as capable as a biological nose would have enormous socioeconomic impact; analogous to the invention of the camera (i.e., the "artificial eye"). It would be transformative for a wide range of industrial and military applications, but, in particular, it would revolutionize health diagnostics.

So what would it take to build a better electronic nose? Electronic noses rely on arrays of chemically distinct materials that each interact uniquely with different gas species, which is closely analogous to how biological noses work. Current electronic noses, however, use small arrays with typically less than 20-30 materials. Dogs, on the other hand, have thousands of distinct olfactory receptors, each of which are replicated tens of thousands of times throughout the interior of their noses. Recent evidence suggests that building larger arrays can significantly improve electronic nose performance, but only if the sensing materials are carefully chosen. For large arrays this becomes a classic "big data" problem, requiring significant computational power and clever algorithms to address the combinatorial explosion of possible arrays one could construct.

In this talk, we discuss recent work on the computational design of gas sensing arrays, based on modeling gas adsorption in metal-organic frameworks (MOFs), and lay out a roadmap for how we might eventually build an electronic nose that could rival a dog's one.

Copyright © American Institute of Chemical Engineers. All rights reserved.

579610 Using Artificial Intelligence to Discover New Materials: The Role of Material Representations

Rational, data-driven materials discovery has the potential to make research and development efforts far faster and cheaper. In such a paradigm, computer models trained to find patterns in massive chemical datasets would rapidly scan compositions and systematically identify attractive candidates. Here, we present several examples of our work on developing machine learning (ML) and deep learning methods capable of creating predictive models using a diverse range of materials data. We illustrate three distinct materials representations and compare the resulting ML models: 1) composition-only attributes, 2) crystal structure-based attributes based on Voronoi tesselation, and 3) an improved version of the crystal graph convolutional neural net (iCGCNN). We also examine the use of "representation-free" approaches based on neural net deep learning. As input training data, we demonstrate ML on both large computational datasets of DFT calculations, as implemented in the Open Quantum Materials Database (oqmd.org), and also experimental databases of materials properties. We construct ML models using a large and chemically diverse list of attributes, which we demonstrate can be used as an effective tool to automatically learn intuitive design rules, predict diverse properties of crystalline and amorphous materials, such as formation energy, specific volume, band gap energy, and glass-forming ability, and accelerate combinatorial searches.

Copyright © American Institute of Chemical Engineers. All rights reserved.

578305 CAR-Mediated T Cell Activation: Linking Molecularly-Detailed and Data-Driven Models

Introduction: Chimeric antigen receptor (CAR) T cells are T cells that have been engineered to express a receptor that binds to a tumor-associated antigen so that the T cells can detect cancer cells better than the normal immune system. While this treatment is successful in many patients, there are also many patients where this treatment is unsuccessful. It has been shown that protein levels can vary between cells of the same type, and that this can lead to phenotypically different cells within the same genotype. Since reaction kinetics depend on many proteins, it is difficult to analyze ODE-based mechanistic models to the fullest extent due to their high dimensionality. Data-driven models, while lacking the biological detail of a mechanistic model, are able to provide a more generalized relationship between model inputs and outputs. Here, we use a data-driven method, in the form of partial least-squares (PLS), to generalize the relationship of protein expression and cell response in CAR T cells. We used the mechanistic model to generate the large amount of data needed for this approach. Thus, the data-driven model is being used as an analysis tool for the mechanistic model. The results of this study should provide better understanding for the improvement of CAR therapy.

Materials and Methods: Using a mechanistic model of CAR T cell signaling, a heterogeneous population of 10,000 cells was simulated. Cell response was determined using the relative phosphorylated ERK concentration. This response separated the population into two categories: responders and non-responders. To better understand the relationship between protein expression and cell response, a PLS model was developed using the initial protein concentrations as inputs and the response as the output. The data generated from the mechanistic model simulations was used to train and validate the accuracy of the PLS model. Examining the components and the weights of the PLS model allowed us to determine the general influence of each individual protein on the cell's ability to respond. Since lower weighted inputs have less of an impact on the PLS outputs, proteins with lower weights were removed from the dataset and the PLS model was retrained to further determine which proteins are highly influential to the cell response.

Results and Discussion: From the population simulations, it was found that only about half of the cells in the population could be characterized as responders. A PLS model was developed that could classify this response with high accuracy. Out of the 19 proteins present in the network, only eight (LAT, Gads, SOS, Ras, RAF, MEK, RasGAP, and RasGRP) were found to be highly influential based upon the weights of their initial concentrations. The influence of these proteins on the system was found to match literature descriptions of their biological roles. Interestingly, it was found that the upstream proteins that interact with the CAR had very little influence on whether or not the cell would respond. All of the influential proteins identified are related to the activation of the MAPK pathway and its signal transduction. This approach of using a data-driven model as a way to analyze a mechanistic model allowed for a generalization of model parameters and investigation of their effects on the system as a whole. Additionally, we have used the data-driven model as a means of easing the computational burden of simulating the full mechanistic model. This enables a multi-scale framework to study immunotherapy, drug resistance, and the effects of tumor cell mutations.

Copyright © American Institute of Chemical Engineers. All rights reserved.

579608 Data Enabled Discovery in Soft Materials Discovery

The advent of new tools for ingestion and analysis of large volumes of data has created new and exciting opportunities for materials discovery and characterization. In the particular case of soft materials, including polymers, rapid advances are taking place on multiple fronts. In this presentation, I will discuss three examples, each of which serves to illustrate a different way in which data science can be used to extract detailed information about molecular processes or interactions. The first is concerned with the role of branching on polymer structure and dynamics, the second is concerned with development of coarse-grained models of conducting polymers, and the third focuses on the use of advanced sampling methods based on the use of concepts from machine learning.

Copyright © American Institute of Chemical Engineers. All rights reserved.

573067 Combining Molecular Dynamics Simulations and Deep Learning to Predict Interfacial Properties

Molecular dynamics (MD) simulations are a powerful tool for analyzing the properties and behavior of materials in solvent environments by explicitly representing atomistic degrees of freedom and accounting for intermolecular interactions. However, MD simulations can require lengthy runtimes that limit the high-throughput screening of system properties. Moreover, the analysis of MD simulation output typically requires the implementation of analytical techniques that rely on human insight to relate the properties of a statistical ensemble to relevant macroscopic properties. Alternatively, machine learning has emerged as a valuable tool to interpret molecular simulation data. Here, we present computational techniques that use machine learning to analyze molecular simulations of complex solvent environments. In particular, we focus on quantifying the hydrophobicity of functionalized interfaces by utilizing deep learning to analyze MD-derived interfacial water configurations. By selecting an appropriate representation of the MD output data, we demonstrate that a convolutional neural network (CNN) can infer spatial correlations between solvent molecules and map these features to thermodynamic observables. Once trained, we find that the CNN can predict dewetting free energies with minimal simulation input, facilitating the rapid screening of interfacial hydrophobicity for a range of surface compositions. We further use feature visualization methods to provide a physical interpretation of features identified by the neural network and compare these features to typical observables used to describe interfacial hydrophobicity (*i.e.*, water structure). This work demonstrates how deep learning methods can complement conventional MD simulations to facilitate the screening of interfacial properties.

Copyright © American Institute of Chemical Engineers. All rights reserved.

579611 Artificial Intelligence in Materials Science: The Good, the Bad, and the Ugly

The revitalization of artificial intelligence (AI) over the last decade due to the dramatic progress in data science has caused resurgence in the development of machine learning methods for materials science. The Holy Grail in materials science is to be able to design and/or discover new materials with desired macroscopic properties rationally, systematically, and quickly, rather than by the slow and expensive trial-and-error Edisonian approach that has dominated for decades. To accomplish this, one has to solve two different but related problems: (i) the forward problem, where one predicts the material properties given the structure or formulation, and (ii) the inverse problem, where one determines the appropriate material structure or formulation given the desired properties. AI methods have the potential to address this challenge in important ways.

However, applying AI to materials science is not new – it has a 35-year history and a rich literature. The forward-inverse conceptual framework, for instance, and its solution using hybrid neural networks and directed evolution, was demonstrated in 1992. What is exciting now is the ability to do all this more easily for more complicated materials due to the availability of powerful and user-friendly hardware and software environments, and, of course, plenty of data.

I classify the materials design problems into three categories – "easy", "hard", and "harder" classes. They correspond to the "good", the "bad", and the "ugly" problems, or the other way around, depending on your persuasion. The relatively "easy" ones are those where there is plenty of data that can be analyzed using off-the-shelf machine learning tools – for example, many structure-to-property prediction problems fall into this class. These can be, and are being, addressed right now. The "hard" problems are those which require combining first-principles knowledge of the underlying physics and/or chemistry with data-driven methods. Although researchers demonstrated how to do this a couple of decades ago, much work still remains to be accomplished to do this systematically and quickly for wider impact. This might take another decade or so.

Harder still is the last class, where one needs to develop domain-specific representations, languages, compilers, ontologies, molecular structure search engines, etc. – i.e., domain-specific "Watson-like" materials discovery engines. These really interesting and intellectually challenging problems would require going beyond purely data-centric machine learning, despite all the current excitement, and leveraging other knowledge representation and reasoning methods from the earlier phases of AI. They would require a proper integration of symbolic reasoning with data- driven processing. This might take a couple of decades to accomplish for extensive and routine usage. I will discuss these challenges and opportunities using materials design examples.

Copyright © American Institute of Chemical Engineers. All rights reserved.

575164 How to Professionally Develop Reusable Scientific Software – and When Not to

A critical challenge in scientific computing is balancing developing high-quality software with the need for immediate scientific progress. We present a flexible approach that emphasizes writing specialized code that is refactored only when future immediate scientific goals demand it. Our lazy refactoring technique, which calls for code with clearly defined interfaces and sharply delimited scopes to maximize reuse and integrability, helps reduce total development time and accelerates the production of scientific results. We offer guidelines for how to implement such code, as well as criteria to aid in the evaluation of existing tools. To demonstrate their application, we showcase the development progression of tools for particle simulations originating from the Glotzer Group at the University of Michigan. We emphasize the evolution of these tools into a loosely integrated software stack of highly reusable software that can be maintained to ensure the long-term stability of established research workflows.

Copyright © American Institute of Chemical Engineers. All rights reserved.

577842 How to Design a Fast Nanocar

How can we design the world's smallest machines—machines that operate at an extraordinarily small scale with molecular moving parts? Just as experimental research on electric motors and combustion engines by nineteenth-century scientists led to revolutionary technologies, exploration of molecular motors, actuators, logic gates, and other molecular machines could lead to similarly transformative discoveries and innovations in the future. However, at present, the focus of computational chemistry software is on exploring various molecular phenomena, not on designing new molecular machines. **Whereas computer-aided design (CAD) software has become mainstream in the design of new cars, airplanes, and other "macro" machines, CAD tools that can design molecular machines have yet to be developed**. Developing comprehensive software for this purpose might seem daunting, even impossible, but a promising starting point is to focus on the design of molecular *components* that can eventually be incorporated into molecular *machines*.

In this work we developed a nanocar builder tool that can be used with a user interface and/or as a Python library for automated building of nanocars. Nanocars are single molecule vehicles that resemble macroscopic automobiles and they are important for understanding how to control molecular diffusion on surfaces. This process is commonly utilized in cells to transport molecular cargo: enzyme molecules are moved along protein filament tracks converting chemical energy into mechanical work. The nanocar builder tool was developed as a plug-in for Avogadro 2 molecular visualizer and a complementary Python package was developed to perform the geometric operations for building the nanocar. The current release allows users to build a nanocar molecule from components of *chassis* and *wheel* molecules. Additionally, users can build a metal surface to place the molecular car and assign force field parameters for the system which provides an initial configuration to study diffusion behavior of these molecules. Using the nanocar builder Python library we generated hypothetical nanocars and studied their diffusion using Molecular Dynamics simulations. We investigated how the molecular interactions effect the diffusion of nanocars and outlined guidelines on how to design faster nanocars.

Currently, the plug-in is available on GitHub (https://github.com/kbsezginel/nanocar-avogadro) including a webpage where users can find detailed written instructions on installation and usage, as well as video tutorials (https://kbsezginel.github.io/nanocar-avogadro). With the upcoming 2nd International Nanocar Race our nanocar builder can be used to improve nanocar designs by understanding how the molecular components of the nanocar affect the diffusion behavior. Additionally, we believe this plug-in serves as a proof-of-concept for a unique platform to build complex molecular machine systems.

Copyright © American Institute of Chemical Engineers. All rights reserved.

575615 Using Unified Memory and Nvlink for Node-Level Strong Scaling of Molecular Dynamics Simulations

In this presentation, I will outline recent additions to the HOOMD-blue simulation toolkit, which leverage the parallelism of dense multi-GPU nodes e.g. of the Summit supercomputer to speed up simulations of complex Molecular Dynamics simulations. HOOMD-blue has been extended to take advantage of CUDA unified memory, a virtual address space that allows coherent access between the CPU and several connected GPUs to it, for optimal performance using the recently introduced NVLINK interconnect. We demonstrate that our implementation can achieve an effective speedup of at least 2x over single GPU execution with up to three GPUs per compute node, for systems that include rigid constraints, pair potentials, and electrostatics. This the first demonstration of the benefit of CUDA unified memory and NVLINK for materials simulation, and we will provide science examples made possible through these capabilities.

Copyright © American Institute of Chemical Engineers. All rights reserved.

561321 Efficient Trajectory Analysis Using Selection Tools in Gromacs

The GROMACS *select* command provides the capability to mine data from trajectories in very powerful ways, especially using the dynamic selection option. However, there is a dearth of information regarding complex program syntax. Program documentation lacks any examples beyond very simple statements. Moreover, error reporting is completely uninformative, complicating its adoption by novice users. Drawing upon prior knowledge and experience with SQL, we have deciphered the GROMACS syntax peculiarities. An example of practical, high throughput, data analysis covering millions of simulation frames will be demonstrated. We foresee useful applications of the *select* command in analyzing solvent-solute interactions, lipid bilayer structures, and the evolution of large structural changes in biomolecular systems. The program allows for simplification of the treatment of periodic boundaries, thus eliminating certain pre-processing steps in data analysis. We have recently integrated the *select* command with other trajectory analysis tools. The command now more seamlessly interfaces with other GROMACS analysis tools, allowing users to mine data that were previously computationally expensive or required external programs. A new tutorial for the implementation of this powerful tool and its applications has been developed to help novice users become masters.

Copyright © American Institute of Chemical Engineers. All rights reserved.

577185 A Software Module for Calculation of Virial Coefficients of Macromolecules

ZENO is a software tool developed at NIST, with capabilities for calculation of properties of polymer and particle suspensions. The latest implementation can compute capacitance, electric polarizability tensor, intrinsic conductivity, volume, gyration tensor, and several more related quantities. Results are obtained by performing appropriate calculations on a single particle instance of the dispersed material, which then enters as the leading order (in concentration f) estimate of the bulk material properties. The dispersed particles may be modeled as arbitrarily shaped objects, including those formed as fused assemblies of smaller objects, such as polyhedra or spheres. The actual particle dispersions modeled this way may be quite complex, with examples including geological systems, food, and blood, as well as engineered systems formulated to improve the mechanical and electrical properties of polymeric materials. A large array of dispersants is of interest, including carbon nanotubes, metal nanoparticles, graphene sheets, polypeptides, and polynucleotides. Significant advances in methods for synthesizing nanoparticles of arbitrary shape open up new possibilities for materials engineering and applications, further driving interest in improving our ability to predict properties for complex suspensions and blends.

Presently, ZENO's capabilities are restricted to single-particle calculations, so its utility can be significantly enhanced by adding features to compute properties using multi-particle metrics. These quantities enter into bulk-property models at second order in concentration, extending the range of conditions where the models are effective. The quantity of interest to this project is the second virial coefficient B2, or more generally the set of coefficients Bn. For particles in solution, these appear in an expansion of the osmotic pressure in terms of the concentration. We consider coefficients with effective interactions mediated by the solvent, so no explicit solvent is considered; rather its effect is manifested by the form chosen for the solute-solute interactions. In principle these interactions are multibody, but often they are treated using pairwise-additive forms. The osmotic virial coefficients can then be effectively given in terms of expressions for the conventional (gas-phase) virial coefficients.

We describe our efforts to implement capabilities within ZENO to compute virial coefficients of arbitrary macromolecules composed of hard spheres. We implement Mayer-sampling Monte Carlo with overlap sampling to reduce the likelihood of bias. The capability builds upon the ZENO framework, and thus may be used with the same input file formats that have already been established by ZENO. Several example applications are demonstrated.

Copyright © American Institute of Chemical Engineers. All rights reserved.

565227 A New Molecular Simulation Code Optimized for the State-of-the-Art Continuous Fractional Component Monte Carlo Method

We present a new molecular simulation code for performing Monte Carlo (MC) simulations in open ensembles using state-of-the-art simulation techniques. In this code, the Continuous Fractional Component (CFC) method is implemented for simulations in the NVT/NPT ensembles, the Gibbs Ensemble, the Grand Canonical Ensemble and the Reaction Ensemble. With the CFC method one can obtain phase equilibria and partial molar properties such as chemical potential, partial molar enthalpy and partial molar volume directly from a simulation without the need of additional simulations or calculations. Furthermore, it is possible to combine MC moves from different ensembles which makes it for example possible to simulate Vapor-Liquid Equilibria in a system where a reaction also takes place. Electrostatic interactions can be calculated with different methods including the Ewald summation and the Wolf method. We demonstrate the applicability of the code by some examples studying industrial relevant processes.

Copyright © American Institute of Chemical Engineers. All rights reserved.

577004 Gomc: GPU Optimized Monte Carlo for the Simulation of Phase Equilibria and Physical Properties of Complex Fluids

GPU Optimized Monte Carlo (GOMC) is open-source software for simulating many-body molecular systems using the Metropolis Monte Carlo algorithm. It supports simulations in a variety of ensembles, which include canonical, isothermal-isobaric, grand canonical, and Gibbs ensemble. This allows GOMC to be used to study vapor-liquid and liquid-liquid equilibria, adsorption in porous materials, surfactant self-assembly, and condensed phase structure for complex molecules. GOMC supports a variety of all-atom, united atom, and coarse grained force fields such as OPLS, TraPPE, Mie, and Martini. The software has been written in object oriented C++, and uses OpenMP and NVIDIA CUDA to allow for execution on multi-core CPU and GPU architectures.

In this work, we discuss recent updates to the GOMC, which include a multi-particle move, parallel tempering, and free energy methods (free energy perturbation and thermodynamic integration), as well as integration with tools for system setup (MoSDeF) and data management (Signac).

GOMC is available from GitHub (GOMC).

Copyright © American Institute of Chemical Engineers. All rights reserved.

572531 Machine Learning for Molecular Property Predictions and the Software Ecosystem That Enables It

The process of creating new chemistry and materials is increasingly driven by computational modeling and simulation, which allow us to characterize compounds of interest before pursuing them in the laboratory. However, traditional physics-based approaches (such as *first-principles* quantum chemistry) tend to be computationally demanding, in which case they may not be a practically viable option for large-scale screening studies that could efficiently explore the vastness of chemical space.

In this presentation, we will show how we employ machine learning to develop data-derived prediction models that are alternatives to physics-based models, and how we utilize them in massive-scale hyperscreening studies at a fraction of the cost. Aside from conducting such data-driven discovery, we also employ data mining techniques to develop an understanding of the hidden structure-property relationships that define the behavior of molecules, materials, and reactions. These insights form our foundation for the rational design and inverse engineering of novel compounds with tailored properties.

In this presentation we will discuss the progress on our software ecosystem for data-driven *in silico* research that enables data-driven research, both on the application as well as on the method development side. It consists of four loosely connected program suites: ***ChemLG*** is a generator for compound and material candidate libraries that allows us to enumerate chemical space (i.e., performing data definition); ***ChemHTPS*** provides an automated platform for the virtual high-throughput screening of these libraries (i.e., performing data generation); ***ChemBDDB*** offers a database and data model template for the massive information volumes created by data-intensive projects (i.e., performing data storage); and ***ChemML*** is a machine learning and informatics toolbox for the validation, analysis, mining, and modeling of such data sets.

The notion to utilize modern data science in chemistry is so recent that much of the basic infrastructure has not yet been developed, or is still in its infancy. The existing tools and expertise tend to be in-house, specialized, or otherwise unavailable to the community at large. Data science is thus in practice beyond the scope and reach of most researchers in the field. By contributing this open, general-purpose, comprehensive, easy-to-use software ecosystem, we aim to chart new paths in this area and help in overcoming this situation, filling the prevalent infrastructure gap, and thus making data-driven research a viable and widely accessible proposition for the community.

Copyright © American Institute of Chemical Engineers. All rights reserved.

577410 Accelerating Large-Scale Screening of Tribological Properties and Chemistries

The tribological properties of monolayer films have been shown to be dependent on the chemistry of the films [1]. This provides a near infinite design space as a means of tuning performance of films. To enable more efficient evaluation of candidate film chemistries, a machine learning model was developed in recent work [2] based on the random forest algorithm. Using chemical descriptors determined from RDKit [3] as the fitting parameters, this model was developed by considering the screening of 420 distinct simulations using the Molecular Simulation and Design Framework (MoSDeF) [4]. Rather than developing a single model, five distinct random forest models were developed using different subsets of the data and performing standard testing; this allowed models that provided poor fitting to be discarded and bounds to be put on predicted values by calculating the mean and standard deviation of the predictions from each of the distinct models. Additionally, adversarial testing was performed on the test data in each model; the differences between testing the entire set of chemical descriptors, subsets of the descriptors that showed the greatest importance, and randomly varying these descriptors, were examined to determine appropriate heuristics for interrogating the models' possible biases. Finally, we performed a series of simulations using chemistries not considered in the original screening to further validate the models. This evaluation suite is being developed as a workflow for the MoSDeF library, allowing best practices for the evaluation of monolayer tribology to be formalized and ultimately disseminated to the community. The overall goal of the work is to allow the a priori identification of films that might possess favorable tribological characteristics, reducing the computational expense required to evaluate the vast landscape of parameters.

1. Brewer, N. J.; Beake, B. D.; Leggett, G. J. Friction Force Microscopy of Self-Assembled Monolayers: Influence of Adsorbate Alkyl Chain Length, Terminal Group Chemistry, and Scan Velocity. *Langmuir* **2001**,*17*(6), 1970–1974.
2. Summers, A.Z.; Gilmer, J.; Iacovella, C.R.; Cummings, P.T.; McCabe, C. Examining Chemistry-Property Relationships in Lubricating Monolayer Films through Molecular Dynamics Screening, *In Preparation*
3. https://www.rdkit.org/
4. http://github.com/mosdef-hub

Copyright © American Institute of Chemical Engineers. All rights reserved.

579415 The Outlook for Physically Informed Machine Learning in Computational Molecular Science

We are witnessing an explosion of interest and activity in the application of artificial intelligence and machine learning areas related to chemical engineering, from determining force fields to materials discovery. We will provide some examples of these applications to look at some studies from "early adopters" to provide a perhaps eclectic snapshot of the current state-of-the-art. This will include some of our own work to apply novel Bayesian Optimization techniques that we have developed to tackle particularly challenging materials discovery problems. These problems include solvent engineering for the solubilization of lead salts in complex organic-inorganic cationic solutions, polymorph predictions in organic semiconductors, and an accelerated route to defect engineering in inorganic electronic materials. These applications are strikingly successful and they inspire confidence in the usefulness of machine learning tools to make dramatic and potentially disruptive changes to materials discovery. However, these studies also uncover some important challenges to their expansion and exploitation. This can occur in the form of traditional and scope-defining chemical engineering concerns, such as scale-up. But the studies also highlight the enduring role of physical insight and intuition and the need to codify these very human traits. The use of these techniques also raises metaphysical questions related to our concept of *understanding* physical processes and our willingness to expand our trust in machine learning-derived predictions that exceed our ability to test them either experimentally or computationally.

Copyright © American Institute of Chemical Engineers. All rights reserved.

557422 Exploiting Electronic Structure and Machine Learning Models for Discovery in Transition Metal Chemistry

Although the discovery and synthesis of new materials, catalysts, and functional molecules represents the foremost effort that unifies thousands of researchers in the chemical science community, presently characterized compounds represent a minute fraction of chemical space. The highly tunable electronic structure properties of inorganic complexes (i.e., variable spin, oxidation state, and coordination number) make them attractive targets for applications in energy storage, functional materials, and catalysis but present a daunting combinatorial challenge. This vast transition metal compound space cannot be fully enumerated by any traditional Edisonian approach. In order to advance quantitative structure-activity relationships, reveal emergent phenomena, and accelerate design of materials and catalysts, smarter and faster computational approaches are needed. In this talk, I will outline our efforts to accelerate first-principles (i.e., with density functional theory, or DFT) screening of inorganic complexes for catalysis and materials science: i) We automate and simplify simulation, eliminating the need for tedious preparation of input files or commands by the user, ii) We develop machine learning (ML) models (e.g., artificial neural networks) that predict the outcomes of simulations prior to their completion, and iii) We integrate these tools into an automated design workflow that can make essential decisions on which simulations are best to carry out and why, with awareness of ML model and DFT model uncertainty. Time permitting, I will describe applications of these tools for advancing understanding in catalysis and functional spin crossover materials.

Copyright © American Institute of Chemical Engineers. All rights reserved.

572554 Thermodynamic and Data-Centric Computational Methods to Guide Nanoparticle Design

Functionalized, monolayer-protected nanoparticles (NPs) are versatile materials with surface properties dictated by the composition of the protecting ligand monolayer. Ultrasmall gold NPs (<10 nm in diameter) are of particular interest for a range of biomedical applications, including targeted drug delivery, biosensing, and bioimaging, because their small size enables unique interactions with serum proteins and the cell membrane. However, structure-property-interaction relationships correlating NP compositions to their biological interactions remain unclear and can be challenging to determine experimentally. NP design is particularly confounded by the large number of design parameters, including core size, ligand physical and chemical properties, and ligand grafting densities, that can influence NP surface properties and thus behavior in biological environments.

Here, we present computational approaches from our group that can guide the design of NPs for biomedical applications. We will first discuss the use of atomistic molecular dynamics simulations to gain physical insight into the effect of specific components of the NP design space on NP surface properties. In particular, we show that ligand fluctuations promoted by the length of the ligand backbone or the NP core size result in spatial variations in surface properties, even for uniform monolayer coatings, and illustrate how these spatial variations influence the magnitude of water-mediated interactions. We then discuss how the specific chemical properties of ligand end groups influence these fluctuations and resulting interactions. To complement these thermodynamic studies, we further apply deep learning to rapidly screen NP surface properties as a function of monolayer composition using minimal simulation input. In conjunction, these thermodynamic and data-centric techniques provide new computational tools that can be leveraged to tune NP surface properties and provide fundamental insight into behavior at nanomaterial interfaces.

Copyright © American Institute of Chemical Engineers. All rights reserved.

576382 Self-Assembly of Complex and Universal Crystal Structures

The ability to create an increasing variety of intricate nanoparticles and colloids has led to the observation of a diverse set of self-assembled structures in the realm of soft materials over the last 15 years [1–5]. While simple, sphere-packing-type arrangements are still commonly found, many of the discovered structures exhibit geometries equivalent to more complex crystal structures on the atomic scale, such as Frank-Kasper phases or clathrates. We want to understand how these building blocks and their self-assembly mechanisms can be engineered to create novel, functional materials.

The formation of different crystal structures can be driven by a variety of particle parameters such as shape, and specific patchy or isotropic interactions [6–8]. Similarly, different atomic compounds crystallize into many of the same crystal structures, despite being driven by different chemical bonding forces, *e.g.*, metallic and van-der-Waals bonding that both cause the formation of close-packed structures, or covalent and hydrogen bonding, either of which cause the formation of the cage-like host structures in silicon-based clathrates *vs.* clathrate hydrates.

To examine these assembly behaviors, we will highlight "universal" crystal structures that occur in a plethora of systems and on various length scales, as well as the necessary conditions under which these crystals form. We will discuss how both direct screening and inverse design approaches explore different aspects of generality and variation in diverse particle self-assembly phase spaces (*e.g.*, [8–11]). Viewing such disparate systems through a common lens gives us insights into designing new types of structures with new geometries and materials properties.

[1] G. Ungar, X. Zeng, Soft Matter 1, 95–106 (2005).
[2] S. Lee, C. Leighton, F. S. Bates, PNAS 111, 17723–17731 (2014).
[3] M. B. Zanjani, J. C. Crocker, T. Sinno, Soft Matter 13, 7098–7105 (2017).
[4] S. Hajiw, B. Pansu, J.-F. Sadoc, ACS Nano 9, 8116–8121 (2015).
[5] H. Lin, S. Lee, L. Sun, M. Spellings, M. Engel, S. C. Glotzer, C. A. Mirkin, Science 355, 931–935 (2017).
[6] P. F. Damasceno, M. Engel, S. C. Glotzer, Science 337, 453–457 (2012).
[7] S. C. Glotzer, M. J. Solomon, Nature Materials 6, 557–562 (2007).
[8] M. Engel, P. F. Damasceno, C. L. Phillips, S. C. Glotzer, Nature Materials 14, 109–116 (2015).
[9] D. Klotsa, E. R. Chen, M. Engel, S. C. Glotzer, Soft Matter 14, 8692–8697 (2018).
[10] Y. Geng, G. van Anders, P. M. Dodd, J. Dshemuchadse, S. C. Glotzer, arXiv:1712.02471, in revision (2019).
[11] C. S. Adorf, J. Antonaglia, J. Dshemuchadse, S. C. Glotzer, J. Chem. Phys. 149, 204102 (2018).

Copyright © American Institute of Chemical Engineers. All rights reserved.

577758 Design of Components and Protocols for Inducing Protective Antibodies Against HIV

No universal vaccines exist for infectious diseases like HIV, flu, and malaria partly because of the high genetic variability of the pathogens that cause these diseases. Vaccination strategies aimed at targeting conserved regions of the pathogenic machinery could mount broadly-neutralizing immune responses, potentially leading to the global eradication of such diseases through a single universal vaccine. Broadly-neutralizing antibodies (bNAbs) have now been isolated from patients with HIV, flu, and malaria that bind to conserved regions on pathogenic proteins called antigens (Ags), and are able to neutralize many different strains of the pathogen. However, the vaccine formulation and administration methods required to optimally promote bNAb formation remains unclear. Through the study of a wide range of possible immunization schemes, this talk will showcase the use of two computational approaches with increasing resolution and biological relevance for studying the development of bNAbs against highly mutable pathogens such as HIV. Results from a heavily coarse-grained model of antibody evolution in a germinal center reaction (GCR) show optimal temporal patterns of administration exist to promote bNAb formation. These optimal temporal immunization schemes can be determined using a predictive framework that has been developed from governing, empirically-determined scaling laws. Results from a more realistic model, focused on studying the sequence evolution and binding of VRC01-like bNAbs to the CD4 binding site of HIV GP120, show that optimal Ags also exist that promote bNAb formation. A two-step, lock-and-key-like mechanism is proposed for bNAb binding to the CD4 binding site.

Copyright © American Institute of Chemical Engineers. All rights reserved.

579414 Toward Engineering the Self-Assembly of Rigid-Soft Multiblock Oligomers

Oligomers and polymers can be tailored to create multi-block oligomers that vary not only in size and chemical composition but also in shape, rigidity, and branching topology. A key challenge that such boundless possibilities present to modelers is the ability to predict the assembling patterns of novel building blocks, and thus potentially identify phases with desirable structures and physical, optical, electronic, catalytic or mechanical properties for emerging applications.

I will describe our work on the phase behavior of multiblock oligomers, i.e., molecules consisting of several chemical block types, focusing on cases where a rigid core is one of the constituent blocks. Our interest centers on architectures that can create complex ordered mesophases that combine solid-like domains interspersed or percolated with liquid-like domains, which can be used, e.g., as dual electron/ionic conducting materials. Our results are filling some of the gaps in the rich phase behavior and the ionic conductivity properties that have been mapped experimentally; indeed, some of our predictions have already been confirmed.

I will also describe some of the newer methodological variants developed in our group to identify suitable order parameters (based on pattern-recognition), to simulate free energies, and to optimize the kinetic behavior for the systems of interest.

Copyright © American Institute of Chemical Engineers. All rights reserved.

578285 Molecular-Scale Experimental and Simulation Synergy to Improve Property Predictions

Experimental validation and support of simulation predictions is not a new scientific pursuit. Collaboration between multidisciplinary teams can produce robust conclusions where vast detailed is provided by models that are strengthened by the experimental backing. Even early applications of classical molecular simulations were benchmarked to experimental data. For example, hard sphere radii can be estimated from liquid structure factors measured directly from scattering experiments, or transferrable forcefields can be tuned to match experimental property measurements (e.g., density, phase equilibria). However, the limits of these approaches should be carefully considered, and overreliance on the accuracy of a simulation developed to model one property should be scrutinized when applied to predict others. In this talk, we will discuss recent experimental and simulation comparisons of the local molecular and nanoscale structure of hydrocarbon systems. Recent construction of neutron scattering equipment designed to measure disordered liquid molecular systems, and accompanying detailed analysis approaches, have opened the opportunity to estimate fully three-dimensional fluid structures. A standard methodology to use these measurements to improve the accuracy of self-assembly predictions has not been developed but serves as potential opportunity. Further the transferability of any such approach to other properties remains uninvestigated.

Copyright © American Institute of Chemical Engineers. All rights reserved.

570358 Multi-Objective, Machine-Learning Assisted First-Principles Design of Transition Metal Complexes for Redox Couples

Truly automatic, in silico design of materials with targeted properties is a central goal in the fields of chemistry, chemical engineering and materials science. This challenge is exacerbated by the enormous number of possible chemistries that could be considered, and by the fact that real materials must fulfil multiple criteria to be useful, e.g. we might seek redox couples that are active, stable and soluble. Unfortunately, many promising materials lie in relatively poorly explored regions of chemical space, for example the space of open-shell transition metal (TM) complexes, which are promising candidates for homogeneous catalysis and molecular electronics. While first-principles simulation with density functional theory (DFT) provides a probe to explore large, heterogenous design spaces, long simulation times and complicated electronic structure severely limits the number of candidates that can be considered. Machine learning (ML) methods can potentially address this problem by using the results of already-complete simulations to generate low-cost surrogate models for target properties, greatly increasing the number of candidates that can be screened. We have previously demonstrated that neural networks based on graph descriptors can be used to predict various properties of TM complexes to near baseline DFT uncertainty with a relatively small number (100s–1000s) of DFT simulations. However, when applying these models to tackle large design spaces, we observed highly variable generalization performance, reflecting the biases and limited variety of chemical motifs in the training data. Therefore, we have introduced geometric metrics to equip our surrogate models with reliable estimates of uncertainty, and we have been able to use these to avoid regions of low data coverage when designing spin crossover complexes and complexes with targeted orbital properties. Here, we use our uncertainty-equipped surrogate models to search a large, million-compound design space for TM complexes that have both target redox energetics and solubility properties for application in redox flow batteries. We use a multi-dimensional formulation of expected improvement that balances exploration (to improve model coverage) and exploitation (to find complexes with targeted properties) to screen this space for lead complexes that we then evaluate with DFT. By repeating this cycle, we can achieve first-principles coverage of the design space only where it is needed, iteratively refining a pareto set of candidate complexes that sample a range of redox and solvation properties.

Copyright © American Institute of Chemical Engineers. All rights reserved.

573946 Machine Learning the Fundamental Tradeoffs between Conductivity and Voltage Stability in Solid State Electrolytes

All solid state batteries provide many safety advantages over traditional lithium-ion batteries by replacing the combustible organic liquid electrolyte with a ceramic solid-state electrolyte. However, reported superionic conductors with conductivities approaching that of liquid electrolytes are unstable in contact with a lithium anode leading to increased internal cell resistance, and poor cyclability. Conversely, compounds stable at the anode or cathode interfaces do not exhibit useful bulk ionic conductivities. Although *ab initio* methods exist to study each ionic conductivity and voltage stability range, there is no established theory to connect these two properties. Here, we leverage machine learning (ML) to investigate the role of crystal structure in the tradeoff between voltage stability and ionic conductivity. To this end, we trained a partial least squares (PLS) machine learning algorithm using the valence electronic density as a descriptor of 60 known solid-state electrolytes along with their corresponding ionic conductivity, anodic voltage limit, and cathodic voltage limit. The trained model has an 80% prediction accuracy and suggests that within the search space of crystal structures, the voltage stability and ionic conductivity are inherently inversely correlated (Figure 1a). A multi-objective optimization also suggests that materials with positively correlated ionic conductivity and voltage stability may be highly anisotropic. Our PLS machine learning model, compared to the more conventional neural network, has the benefit of being able to predict and explore the relationship between multiple properties and retains a high level of interpretability versus other 'black box' models. The PLS model successfully identifies BCC anion substructure (Figure 1b) and channels as effective descriptors, which is in good agreement with prior work.[1] Using this model, we screened through a database of ca. 14,000 materials and identified five new promising solid state electrolyte candidates to have conductivities greater than 16 mS/cm. The model predictions were subsequently verified with *ab initio* molecular dynamics simulations. The proposed ML model and electron density descriptor may be used in future studies to elucidate structure-property relationships for other applications with high accuracy and without sacrificing interpretability.

Figure 1: (a) Latent variable space determined by the PLS model used to predict properties and (b) the BCC anion substructure identified by the PLS model.

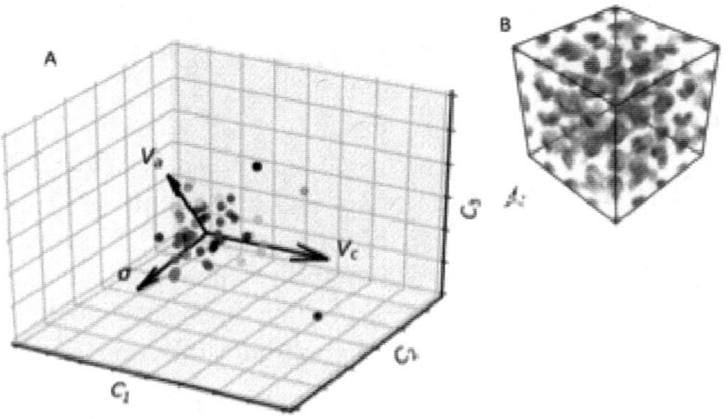

References:

1. Wang, Y. *et al.* Design principles for solid-state lithium superionic conductors. *Nat. Mater.* **14,** 1026–1031 (2015).

Copyright © American Institute of Chemical Engineers. All rights reserved.

576672 Machine Learning for the Prediction of Electronic Couplings and Charge Transport Calculations in P3HT

Predicting which organic chemical blends will make desirable solar cell materials requires a suite of computational chemistry tools including molecular dynamics simulations, quantum chemical calculations, and kinetic Monte Carlo simulations. In this work, we focus on ameliorating a quantum chemical calculation bottleneck in the pipeline between chemical ingredients and the charge mobilities that results from the self-assembled structure in these soft materials. Here we focus on the benchmark polymer poly(3-hexylthiophene) (P3HT) and calculating the dependence of charge mobility on P3HT nanostructure with kinetic Monte Carlo simulations that take charge hopping rates between chromophores as inputs. We compare five machine learning techniques for correlating structural features of P3HT monomers with the electronic couplings calculated with Zerner's method of intermediate neglect of differential overlap for spectroscopy (ZINDO/S). We find that random forests are an easy-to-implement, accurate predictor of couplings that finds the separation and orientation of bonded and non-bonded P3HT monomers chiefly determine their coupling. We discuss tests informing the minimal training set needed to generate random forests with 0.02 mean absolute error that is 400 times faster than performing new ZINDO/S calculations. We find that the charge mobilities predicted with the random-forest-generated couplings to be in agreement with the ZINDO/S mobilities. We conclude with an overview of the pipeline in which these random forests are used and how they are enabling broader screening of organic photovoltaic candidate chemistries.

Copyright © American Institute of Chemical Engineers. All rights reserved.

564718 Endotoxin Sensors Using Liquid Crystals and Machine Learning

Endotoxins are lipopolysaccharides that are found in the outer membranes of gram-negative bacteria and underlie many health hazards such as endotoxemia (a septic shock caused severe immune response) [1]. A number of methods have been proposed to detect the presence of endotoxins with remarkable sensitivity [2,3,4]. However, the most commonly used methods rely on complex reagents derived from the blood of crabs, do not provide a quantitative estimation of concentration, and/or are incapable of classifying endotoxin type.

Liquid crystal (LC) droplets dispersed in water are sensitive to the presence of endotoxins and undergo internal ordering transitions when exposed to different endotoxin species and concentrations [5]. Specifically, endotoxin accumulates at topological defects within the LC droplets, triggering a change from droplet configurations with bipolar symmetry to droplet configurations with radial symmetry. Different internal configurations of LC droplets have distinct optical properties [6]. As such, flow cytometry of LC droplets can be used to generate rich forward scattering/side scattering (FSC/SSC) data. A variety of counting methods used in the analysis of FSC/SSC data can predict endotoxin concentration [7], however, the optimal method for characterizing the scatter plots is not known and some methods are subject to variability.

In this work, we demonstrate that machine learning techniques can be used to accurately predict endotoxin concentrations and species from FSCC/SSC data. Our approach uses a convolutional neural network to extract pattern information (features) from FSC/SSC density data. Our framework reveals that significant hidden information is available in FSC/SSC data. Our analysis also provides insights into the physical effects driving changes in the LC internal configurations.

References:

[1] R. A. Proctor, *Handbook of Endotoxin*. Elsevier, 1984.

[2] T. Muta, T. Oda, and S. Iwanaga, "Horseshoe crab coagulation factor B. A unique serine protease zymogen activated by cleavage of an Ile-Ile bond.," *Journal of Biological Chemistry*, vol. 268, no. 28, pp. 21384–21388, 1993.

[3] K. G. Ong, J. M. Leland, K. Zeng, G. Barrett, M. Zourob, and C. A. Grimes, "A rapid highly-sensitive endotoxin detection system," *Biosensors and Bioelectronics*, vol. 21, no. 12, pp. 2270–2274, 2006.

[4] T. Y. Yeo *et al.*, "Electrochemical endotoxin sensors based on TLR4/MD-2 complexes immobilized on gold electrodes," *Biosensors and Bioelectronics*, vol. 28, no. 1, pp. 139–145, 2011.

[5] N. Bao, Y. Zhan, and C. Lu, "Microfluidic Electroporative Flow Cytometry for Studying Single-Cell Biomechanics," *Analytical Chemistry*, vol. 80, no. 20, pp. 7714–7719, 2008.

[6] I.-H. Lin, D. S. Miller, P. J. Bertics, C. J. Murphy, J. J. de Pablo, and N. L. Abbott, "Endotoxin-Induced Structural Transformations in Liquid Crystalline Droplets," *Science*, vol. 332, no. 6035, p. 1297, Jun. 2011.

[7] D. S. Miller, X. Wang, J. Buchen, O. D. Lavrentovich, and N. L. Abbott, "Analysis of the Internal Configurations of Droplets of Liquid Crystal Using Flow Cytometry," *Anal. Chem.*, vol. 85, no. 21, pp. 10296–10303, Nov. 2013.

Copyright © American Institute of Chemical Engineers. All rights reserved.

559792 Microscopic Modeling and Optimal Operation of Thermal Atomic Layer Deposition

Atomic layer deposition (ALD) process has been widely adopted by industry to produce thin-film materials that meet the requirements of major design breakthroughs in semiconductor industry [1]. Despite its popularity, the ALD process remains costly and it is desirable to develop an inexpensive high-fidelity model to characterize the process [2]. Recently, Shirazi and Elliot developed a Density Functional Theory (DFT)-based simulation of the HfO_2 deposition on laboratory scale [3]. Motivated by this work, a similar approach is expanded to develop a comprehensive framework for industrial-scale first-principles-based microscopic modeling, data-driven input/output modeling and optimal operation of thermal ALD of SiO_2 thin-films using bis(tertiary-butylamino)silane (BTBAS) and ozone as precursors [4].

In this work, we first perform first-principles-based DFT calculations of the key thermodynamic and kinetic parameters, which are then used in the microscopic modeling of the ALD process. Subsequently, a detailed microscopic model is constructed, accounting for the microscopic lattice structure and atomic interactions, as well as multiple microscopic film growth processes including physisorption, abstraction and competing chemical reaction pathways. Kinetic Monte-Carlo (kMC) algorithms are utilized to obtain computationally efficient microscopic model solutions while preserving model fidelity. The obtained kMC simulation results are used to train Artificial Neural Network (ANN)-based data-driven models that capture the relationship between operating process parameters and time to ALD cycle completion. Specifically, a dense two-hidden-layer feed-forward ANN is constructed to find a feasible range of ALD operating conditions accounting for industrial considerations, and a Bayesian Regularized ANN is constructed to implement the cycle-to-cycle optimization of ALD cycle time. Extensive simulation results demonstrate the effectiveness of the proposed approaches. The kMC models successfully achieves a growth per cycle (GPC) of 1.8 Å per cycle, which is in the range of reported experimental values. The ANN models accurately predict deposition time to steady-state from the given operating condition input, and the cycle time optimization using BRANN model reduces the conventional BTBAS cycle time by 60%.

Copyright © American Institute of Chemical Engineers. All rights reserved.

563078 Thermodynamic Properties Via a Combined Expanded Wang-Landau - Machine Learning Approach

Using machine learning (ML), we predict the partition functions and, thus, all thermodynamic properties of atomic and molecular fluids over a wide range of temperatures and pressures. Our approach is based on training neural networks using, as a reference, the results of a few flat-histogram simulations. The neural network weights so obtained are then used to predict fluid properties that are shown to be in excellent agreement with the experiment and with simulation results previously obtained on argon, carbon dioxide, and water. In particular, the ML predictions for the Gibbs free energy, Helmholtz free energy, and entropy are shown to be highly accurate over a wide range of conditions and states for bulk phases as well as for the conditions of phase coexistence. Our ML approach thus provides access instantly to G, A, and S, thereby eliminating the need to carry out any additional simulations to explore the dependence of the fluid properties on the conditions of temperature and pressure. This is of particular interest, for e.g., the screening of new materials, as well as in the parameterization of force fields, for which this ML approach provides a rapid way to assess the impact of new sets of parameters on the system properties. We then develop a combined Expanded Wang-Landau-Machine Learning (EWL-ML) approach to predict efficiently and accurately the thermodynamics of mixtures. The resulting speed-up is expected to considerably increase the range and complexity of systems that can be studied with even-sampling methods.

Copyright © American Institute of Chemical Engineers. All rights reserved.

569671 Understanding Mixing Behavior of Bimetallic Nanoparticles through Genetic Algorithm Modeling

Bimetallic nanoparticles (NPs) have received tremendous interest due to their controllable properties that make them promising candidates for a wide range of applications, such as catalysis and electronics. Central to these applications is to predict the thermodynamic preference of metals mixing at the nanoscale as a function of NP size, shape and metal composition. However, a complexity in bimetallic NPs is the explosion in the size of configurational space. In fact, it becomes impossible to study the entire search space of bimetallic NPs as their size increases, especially if one relies on first-principles calculations. To combat this problem, we develop a genetic algorithm (GA) capable of targeted screening towards the discovery of stable bimetallic NPs of any given morphology. Our GA leverages a previously reported "Bond-Centric" cohesive energy model[1] as a fast and accurate stability metric with practically minimal computational cost. With this approach we achieve high-throughput screening and can effectively search the large configurational space of bimetallic NPs. Using results from our GA, we connect the morphology, composition, and chemical ordering to the mixing behavior of bimetallic NPs – focusing on combinations of Cu, Ag, and Au. Importantly, we incorporate entropic effects into our framework to determine how mixing behavior changes with temperature. Taken together, our results unravel the impact of NP morphology (size, shape), metal composition, and temperature to the thermodynamic mixing behavior of bimetallic NPs and elucidates their detailed structural architectures (i.e. positioning of different metals on different NP sites).

1. Yan, Z.; Taylor, M. G.; Mascareno, A.; Mpourmpakis, G., Size-, Shape-, and Composition-Dependent Model for Metal Nanoparticle Stability Prediction. *Nano Lett.* **2018,** *18* (4), 2696-2704.

Copyright © American Institute of Chemical Engineers. All rights reserved.

573708 Estimation of Probability Density Functions of Model Parameters for a Heterogeneous Population with Neural Network: Application to TNFα Signaling Pathway

Modeling a heterogeneous population, whose dynamics are governed by same differential equations but with different parameter' values, has received much attention in various disciplines such as quantum physics and cellular biology [1-3]. Previous studies mostly focused on the problem of the optimal control of this heterogeneous population while methods on how to estimate such systems have begun to emerge [3-6].

Motivated by this, this study proposes a methodology to develop a heterogeneous dynamic model for a heterogeneous population. Specifically, we propose a numerical method to estimate the probability density functions (PDFs) of model parameters from experimental measurements. Here, we particularly consider the case where the output PDFs are available in the form of population snapshot data (PSD): PSD provide measurements from all members in the population at each time instance [6]. First, the PDFs of the model parameters are assumed to be normal so that the overall estimation problem become finite-dimensional. Second, Sobol' sensitivity analysis method is performed to determine a set of model parameters whose means and variances are practically identifiable. Third, a neural network model is developed to find an empirical function mapping from the PDFs of the identifiable parameters to the output PDFs. Lastly, the PDFs of the identifiable parameters are estimated by solving an optimization problem to minimize the difference between the measured and predicted output PDFs, which are computed based on PSD and the developed neural network model, respectively.

Our proposed methodology was implemented to infer the PDFs of the parameters in a tumor necrosis factor-α (TNFα) signaling pathway. Artificial PSD of the TNFα signaling dynamics were generated by simulating the model accordingly. Next, the proposed method was implemented to determine identifiable parameters and infer their PDFs from the PSD. After inferring the PDFs of these identifiable parameters, the corresponding output PDFs were calculated and compared with the PSD to validate the inference results. The reasonable agreement between the predicted and measured output PDFs demonstrated the validity of the proposed methods.

References

[1] Li, J.-S.; Khaneja, N. Control of inhomogeneous quantum ensembles. *Physical Review A* **2006**, *73*, 030302.

[2] Waldherr, S. Estimation methods for heterogeneous cell population models in systems biology. *Journal of the Royal Society Interface* **2018**, *15*, 20180530.

[3] Zeng, S.; Waldherr, S.; Allgöwer, F. Ensemble Observability of Linear Systems. *IEEE Transactions on Automatic Control* **2016**, 61, 1452-1465.

[4] Li, J.-S. Ensemble control of finite-dimensional time-varying linear systems. *IEEE Transactions on Automatic Control* **2011**, *56*, 345-357.

[5] Helmke, U.; Schönlein, M. Uniform ensemble controllability for one-parameter families of time-invariant linear systems. *System & Control Letters* **2014**, *71*, 69-77.

[6] Hasenauer, J.; Waldherr, S.; Doszczak, M.;Scheurich, P.; Radde, N.; Allgöwer, F. Analysis of heterogeneous cell populations: A density-based modeling and identification framework. *Journal of Process Control* **2011**, *21*, 1417-1425.

Copyright © American Institute of Chemical Engineers. All rights reserved.

567628 Development of Multimodal Cancer Diagnosis Method Based on Medical Image Analysis and Raman Spectroscopy

Digital pathology (DP) [1] is playing an ever-increasing role in clinical decision making. Technological advancements have made efficient data recording, processing, and analyzing possible. This enables the transformation of cancer diagnosis such that diagnosis can now be routinely assisted by algorithms in, for instance, computational image analysis [2], not least a wide variety of artificial neural networks (ANNs) developed for computer vision. Meanwhile, effective (and perhaps more intuitive) cancer identification and classification can also be realized based on appropriately selected morphological features of the biopsy samples, e.g. object contour, color, and texture [1]. It is therefore desirable to combine the advantages of both approaches so that a reliable and easy-to-interpret diagnosis can be achieved. Moreover, Raman spectroscopic analysis of specimens is gaining popularity in the biomedical research community [3], as molecular composition and microenvironment information of the cells unavailable in medical images can be highly sensitively captured in the Raman spectroscopy. Thus, Raman spectral profiles and their spatial distribution can be used to assist in characterizing cell/tissue samples. In this work, we construct a multimodal cancer diagnosis model based on the information provided by both the medical image analysis and Raman spectroscopy. Furthermore, we demonstrate the effectiveness and efficiency of the approach by running it on a wide variety of diagnostic samples.

[1] R. Bhargava and A. Madabhushi, *Annu. Rev. Biomed. Eng.* **18**, 387–412 (2016).

[2] D. Shen, G. Wu, and H.-I. Suk, *Annu. Rev. Biomed. Eng.* **19**, 221–248 (2017).

[3] H. J. Butler *et al.*, *Nat Protoc.* **11**, 664-687 (2016).

Copyright © American Institute of Chemical Engineers. All rights reserved.

576578 Time-Independent Free Energy Surfaces from Time-Dependent Biased Simulations Via Thermodynamic Integration

Free energy surfaces, i.e. projections of the free energy of a system in a low-dimensional set of collective variables, are often employed to convey quantitative information on the thermodynamics of molecular processes[1].

History-dependent biased sampling methods, such as metadynamics[2,3], provide a convenient way of adaptively exploring the configuration space of a molecular system. However, in order to efficiently leverage the sampling obtained from such protocols, time-independent estimates of free energy surfaces are often desirable[4,5].

Here, inspired by thermodynamic integration[6], we propose a general method for the calculation of time-independent free energy surfaces from history-dependent biased simulations. In contrast with other methods, our proposed approach is not limited to specific sampling protocols, for instance we demonstrate its applicability on different variants of metadynamics including standard, Well Tempered, and Adaptive Gaussians.

In all cases the method displays faster convergence compared to conventional estimates, in which free energy surfaces are computed as a function of the total bias in the long-time limit. Moreover, our approach circumvents the need for computing the time-dependent constant $c(t)$ as done with the weighted histogram analysis method (WHAM)[7] or through the reweighing procedure proposed by Tiwary and Parrinello[5].

Remarkably, the proposed method extends naturally to aggregate information obtained from independent simulations carried out with different biasing protocols in a single estimate of the free energy surface, allowing us to efficiently compute free energy surfaces from swarms of independent metadynamics runs.We note that, in order to achieve convergence with our method, while sampling barrier crossing events is necessary, it is not compulsory to observe re-crossing events in a single continuous trajectory. This feature allows to efficiently estimate and refine free energy surfaces with short, independent calculations.

In this work we validate our approach against one and two-dimensional model potentials, and by comparing the conformational free energy surface of ibuprofen in the solid state[8] computed from a single, long well-tempered metadynamics simulation and from a swarm of short, independent metadynamics runs.

Finally, we discuss the application of our proposed method to efficiently estimate the reversible work necessary to form a kink site on a series of ibuprofen crystal surfaces in different solvents.

References

1. Chipot, C., & Pohorille, A. (**2007**). *Free energy calculations*. Springer-Verlag Berlin Heidelberg.
2. Laio, A. and M. Parrinello. *Proceedings of the National Academy of Sciences* 99.20 (**2002**): 12562-12566.
3. Barducci, A., G. Bussi, and M. Parrinello. *Physical review letters* 100.2 (**2008**): 020603.
4. Bonomi, M., A. Barducci, and M. Parrinello. *Journal of computational chemistry* 30.11 (**2009**): 1615-1621.
5. Tiwary, P., and M. Parrinello. *The Journal of Physical Chemistry B* 119.3 (**2014**): 736-742.
6. Kästner, Johannes, and Walter Thiel. *The Journal of chemical physics* 123.14 (**2005**): 144104.
7. Kumar, S., Rosenberg, J. M., Bouzida, D., Swendsen, R. H., & Kollman, P. A. (**1992**). *Journal of computational chemistry*, *13*(8), 1011-1021.
8. Marinova, V., G. Wood, I. Marziano, M. Salvalaglio, *Journal of chemical theory and computation* 14.12 (**2018**): 6484-6494.

Copyright © American Institute of Chemical Engineers. All rights reserved.

572760 Mapping Saddles and Minima on Free Energy Surfaces Using Multiple Climbing Strings

Locating saddle points on high dimensional free energy surfaces is key in characterizing multistate transition events and associated transition rates in complicated molecular-scale systems. Though finding minima is a routine task, locating saddles and measuring their energies relative to their associated minima remain challenging, especially in high-dimensional spaces. We propose here a multistring version of climbing string method to locate multiple saddles in collective variable space by initiating multiple strings in one go. The multistring method employs dynamic strings to locate saddles which propagate by the gradient flow in the path space, with one end fixed and other end climbing across the free energy profile. The method further introduces the static strings to keep the history of prior strings converged to saddles. The inter-string interactions are used to avoid the convergence of multiple strings to a common saddle. The presence of multiple images along the string helps to ensure the direct connectivity between the minimum and saddles and to compute the free energy along the path. Hence, the climbing multistring method can be used to map the network of directly connected stationary points and their associated activation barrier. We demonstrate this method to locate saddle points in two-dimensional and four-dimensional free energy surface of alanine dipeptide and alanine tripeptide, respectively.

References:

1. Ren, W.; Vanden-Eijnden, E. J. Chem. Phys. 2013, 138, 134105

2. Maragliano, L.; Vanden-Eijnden, E. Chem. Phys. Lett. 2006, 426, 168–175.

*Funding for this project was provided by NIH grant number GM 100472.

Copyright © American Institute of Chemical Engineers. All rights reserved.

571166 The Overdamped Transmission Coefficient: Recovering the True Mean First Passage Time from an Inaccurate Reaction Coordinate

For inertial reaction dynamics, a transition state theory rate constant obtained from an inaccurate reaction coordinate can be a posteriori corrected with reactive flux methods. In contrast, reaction coordinate errors in overdamped mean first passage time calculations cannot be a posteriori corrected. This work develops an overdamped version of the transmission coefficient. The calculation requires information from committor analyses and an estimate of the diffusivity along the committor coordinate. We illustrate the calculation for a simple two-dimensional potential that admits exact solutions.

Copyright © American Institute of Chemical Engineers. All rights reserved.

574370 Collective Mode Brownian Dynamics: A Method for Fast Relaxation of Statistical Ensembles

Sampling equilibrium configurations of correlated systems of particles with slow relaxation times (e.g., polymeric systems) using conventional molecular dynamics and Monte Carlo methods can be challenging. This is especially true for systems with complicated, extended bond network topologies and other interactions that make the use and design of specialized relaxation protocols infeasible. We introduce a method based on Brownian dynamics (BD) simulations that can reduce the computational time it takes to reach equilibrium and measure decorrelated samples. Importantly, the method is completely agnostic to particle configuration and the specifics of interparticle forces. In particular, we develop a mobility matrix that excites non-local, collective motion of N particles and can be computed efficiently in $O(N)$ time. Particle motion is integrated according to the overdamped Langevin equation with hydrodynamic interactions, where Brownian displacements are drawn efficiently using a split representation of the mobility matrix in position and wave space. We demonstrate the efficacy of the method with various examples from the realm of soft condensed matter.

Copyright © American Institute of Chemical Engineers. All rights reserved.

568426 A New High-Throughput Computational Method to Generate Energy Profile of Integral Membrane Protein-Protein Association

Integral membrane proteins play a key role in crucial biological events including neurotransmission, signaling cascade, immune response, and cancer metastasis among others. Not surprisingly, these membrane proteins represent 20-30% of the proteome of most organisms. Despite their significance, isolation and characterization of membrane proteins is experimentally challenging due to their hydrophobic membrane environment. Several computational (docking and self-assembly) methods to study protein assembly have been developed but have limited scope or affordability. There is growing need for a comprehensive and affordable method that captures molecular-level detail of protein association in complex lipid environments. Here we report the development of a new method that provides qualitative and quantitative estimates of membrane protein association to generate a comprehensive Protein Association Energy Landscape (PANEL). The PANEL method is a computationally inexpensive approach to generating energy profiles for a wide range of proteins belonging to diverse set of species in variable membrane environments. We have used our approach to show association of membrane proteins that are responsible for the selectivity of the blood-brain tight junctions.

Copyright © American Institute of Chemical Engineers. All rights reserved.

567359 Development of a New Coarse-Grained Model Capturing Hydrogen Bonding Effects in Polysaccharides

Novel applications of cellulose require a molecular-level understanding of its crystalline structure and its inter-chain hydrogen bonding network. Molecular simulation is a valuable tool to understand the microscopic interactions and the resulting structure in biomacromolecules. While atomistic simulations are chemically detailed enough to capture the hydrogen bonds, they do not capture the large time and length scales of the macromolecular systems like cellulose. Simulations using coarse-grained models have been able to investigate cellulose at macromolecular length and time scales but do not capture the directional nature of the hydrogen bonding interactions. In this talk, we present a new coarse-grained model of cellulose that is able to capture the directional nature of hydrogen bonding in cellulose. Simulations performed with this model demonstrate the aggregation and assembly of cellulose chains into sheet-like parallel structures. We investigate the inter-chain hydrogen bonding pattern within the aggregate and compare our results with experimentally-established cellulose crystalline structures. We also explore the potential of this model to represent modified and substituted cellulose and examine how the assembly structure can be tuned by these substitutions.

Copyright © American Institute of Chemical Engineers. All rights reserved.

568783 Implementation of Quantum Mechanics/Coarse-Grained Molecular Mechanics (QM/CG-MM)

Advances in understanding of complex biochemical processes using molecular simulations critically depend on the availability of accurate and computationally efficient force fields, enabling efficient exploration of the configurational space. As the computational cost of classical all-atom (AA) force fields becomes increasingly high and insurmountable for certain classes of proteins (e.g., actin) or lipid membranes, bottom-up coarse-graining (CG) models offer a viable alternative. The multiscale coarse-graining methodology (MSCG),[1] based on the variational principle of force matching, enables access to longer time and larger length scales through the reduction of the number of degrees of freedom with minimal loss of accuracy, as has been demonstrated for a disparate set of systems ranging from ionic liquids to lipids.

Current classical AA and CG methods are inherently limited to processes that preserve the topology of molecules involved, and thus cannot describe chemical bond breaking and formation. This limitation has been partially lifted with the advent of the Quantum Mechanics/Molecular Mechanics (QM/MM) method, which treats a small region undergoing chemical transformations quantum mechanically, whereas the rest of the system is described classically. However, the large number of MM degrees of freedom together with the high computational cost of QM energy evaluations render this method unsuitable for large protein molecules.

Sinitskiy and Voth[2] proposed the QM/CG-MM theory, which holds promise to surmount QM/MM limitations by coarse-graining the MM part. In this work, we evaluate one implementation of this method based on the MSCG theory. We consider two model systems: QM CCl_4 embedded in (CG-) MM CCl_4 (non-reactive non-polar system) and the reaction of tert-butyl hypochlorite with benzyl radical in (CG-)MM CCl_4, for which the experimentally measured reaction rate constant is available[3] to determine the method's accuracy. We assess the method's performance relative to QM/MM by comparing radial distribution functions and the potential of mean force. Finally, we discuss strategies to extend the method to polar solvents.

References:

1. Izvekov, S., & Voth, G. A. (2005). A multiscale coarse-graining method for biomolecular systems. *The Journal of Physical Chemistry B*, 109(7), 2469-2473.
2. Sinitskiy, A. V., & Voth, G. A. (2018). Quantum mechanics/coarse-grained molecular mechanics (QM/CG-MM). *The Journal of chemical physics*, 148(1), 014102.
3. Zavitsas, A. A., & Blank John, D. (1972). Kinetics of the free-radical chain chlorination of hydrocarbons by tert-butyl hypochlorite. *Journal of the American Chemical Society*, 94(13), 4603-4608.

Copyright © American Institute of Chemical Engineers. All rights reserved.

565099 An Alternative Model for Simulating Water between Two Phosphorylcholine Self-Assembled Monolayer Surfaces

The properties of water in confined systems have been an interesting topic of research for years. The understanding of these properties is vital to many areas such as tribology and microelectromechanical systems. Yet, it is very challenging to implement a cost-effective model to study water in confined systems with molecular dynamics simulations because these models typically require the use of grand-canonical (GC) simulation or thermodynamic extrapolation methods (TEM). In this work, we proposed and examined an alternative but effective model for the study of water in confined systems using the standard Large-scale Atomic/Molecular Massively Parallel Simulator (LAMMPS) without any additional customized subroutines. Two pseudo surfaces were used to play the role of pistons for maintaining pressure and two phosphorylcholine self-assembled monolayers (PC-SAM) were used as model surfaces to implement the confinement. Both the structure and dynamics of the water was investigated. In addition, the model was used to study the friction between the two surfaces, producing simulation results comparable to what has been reported in the literature. The calculated pressure of water between the surfaces was also in good agreement with experimental values. Therefore, the model may provide a cost-effective way to explore the properties of water in confined systems at the molecular level. In principle, the model should be applicable for other fluids in other confined systems as well.

Copyright © American Institute of Chemical Engineers. All rights reserved.

578289 Van Hove Correlations in Aqueous Systems: Insights from Molecular Simulation

Van Hove Correlations in Aqueous Systems: Insights from Molecular Simulation

Matthew W. Thompson[1,2], Ray Matsumoto[1,2], Peter T. Cummings[1,2]

[1]*Multiscale Modeling and Simulation Center, Vanderbilt University, Nashville, TN, 37235-1604, USA*

[2]*Department of Chemical and Biomolecular Engineering, Vanderbilt University, Nashville, TN, 37235-1604, USA*

The physical properties of water and aqueous solutions are of ubiquitous and self-evident interest to innumerable fields. Molecular simulation has for several decades informed experimental advances in describing the microscopic behavior of these molecules and how such interactions affect macroscopic properties. It is therefore of continued interest to evaluate the accuracy of different molecular simulation methods in describing these properties in exhaustive detail. Most of these properties are understood as either static structures computed via ensemble averages or transport properties computed directly or indirectly at timescales longer than the rotational relaxation time of most water models. Recent advances in neutron and X-ray scattering experiments have made it feasible to study the Van Hove correlation of water molecules and some aqueous electrolytes, which make it possible to study the time-dependent structure of these liquids at very fine resolutions, on the order of less than a picosecond in time and less than an angstrom in distance. A major limitation of these experiments, however, is that the signal produced is a statistical average over all components of a system; therefore it is difficult to understand from experiments alone which atomic interactions most strongly drive molecular behavior. Here we present results from a series of studies aimed both at better understanding these physical systems and the robustness of different force fields and molecular simulation techniques at describing them. Included are results from a broad set of force fields, encompassing classical, polarizable, reactive, and first-principles molecular dynamics simulations. Also included is an investigation of how the sizes of ions in aqueous solutions impact the very short-time and short-range correlated motions and how these observations clash with some conventional understandings of the transport properties of these solutions.

Iwashita, T., Wu, B., Chen, W.-R., Tsutsui, S., Baron, A. Q. R., & Egami, T. (2017). Seeing real-space dynamics of liquid water through inelastic x-ray scattering. *Science Advances*, *3*(12), e1603079. https://doi.org/10.1126/sciadv.1603079

Van Hove, L. (1954). Correlations in space and time and born approximation scattering in systems of interacting particles. *Physical Review*, *95*(1), 249–262. https://doi.org/10.1103/PhysRev.95.249

Marcus, Y. (1988). Ionic radii in aqueous solutions. *Chemical Reviews*, *88*(8), 1475–1498. https://doi.org/10.1021/cr00090a003

Copyright © American Institute of Chemical Engineers. All rights reserved.

573617 Quantification of DNA Hybridization Thermodynamics at the Atomistic Level

It is widely accepted that DNA hybridization follows a classic nucleation-zippering mechanism. Sampling the nucleation barrier via conventional MD is notoriously difficult and hence inaccurate. Commonly used advanced sampling approaches are also of limited applicability for high-resolution, all-atom studies of DNA hybridization. Accordingly, several coarse-grained models of DNA have been developed to accurately quantify DNA hybridization thermodynamics. However, for some fundamental questions, such as the influence of base chirality on hybridization, atomistic resolution is required. We describe an order parameter-based advanced-sampling technique to compute the hybridization free energy and melting temperature of DNA oligomers using a recently developed native topology-based order metric. We illustrate the implementation of this technique in a study of chiral perturbations to a DNA oligomer and how effect on hybridization thermodynamics.

Copyright © American Institute of Chemical Engineers. All rights reserved.

560000 Impact of Uncertainty Quantification and Density Functional Theory Error Propagation on Surface Catalyzed Kinetic Models

Parametric uncertainty currently limits the predictive ability of kinetic models and prevents them from reaching their full potential. This is particularly important for surface catalyzed microkinetic models which often contain thousands of reaction steps and intermediate surface species requiring many kinetic and thermodynamic parameters. Estimating these from first-principles density functional theory (DFT) is costly, motivating the need for DFT-based methods, such as group additivity and Brønsted–Evans–Polanyi (BEP) relationships, to quickly estimate thermochemistry and reaction barriers, respectively. Uncertainty in DFT and these methods is amplified by the Boltzmann factors in the reaction rate expressions increasing uncertainty in estimating the overall reaction rate. Furthermore, microkinetic model parameters are correlated; yet, there is lack of a framework to expose these correlations on a single catalyst with a single functional. This work develops a framework to estimate correlations in reaction networks and DFT uncertainty propagation. Uncertainty quantification (UQ) is then predicted for thermochemistry, reaction barriers, reaction path, and ultimately reaction rates. Propane combustion and ethane oxidative dehydrogenation reactions are illustrated. We will also demonstrate how to use the Python Multiscale Thermochemistry Toolbox, pMuTT [1], to calculate uncertainty in apparent activation energy.

References

[1] Python Multiscale Thermochemistry Toolbox (pMuTT), https://pypi.org/project/pMuTT/

Copyright © American Institute of Chemical Engineers. All rights reserved.

568912 Automated Machine Learning and Microfluidics-Based Study and Optimization of Zeigler-Natta Catalysts

Development of new polymerization processes and the requisite catalysts is a complex task, often taking years and the combined efforts of multiple people, especially when optimizing for various parameters such as optimal chain length, polydispersity index, and economic factors. This challenge has only increased in recent years as there has been a move toward homogeneous Ziegler-Natta catalysts, a class of molecules which provides favorable control and processing characteristics but are notoriously difficult to optimize and understand. Optimization of a Ziegler-Natta catalyst system relies on the careful selection of several key parameters including the solvent, type and concentration of activator, catalyst concentration, monomer addition, temperature and mixing. While all these factors play into generating the desired polymer, the relationship between these variables is often not clear due to the chemistry involved. Proposed mechanisms for this reaction include up to eleven different equilibrium and non-equilibrium reaction steps, the careful balance of which effects the properties of the final polymer. Traditionally this has been reconciled by performing combinatorial experiments to individually judge the contribution of each factor, but these experiments take time and create large amounts of chemical waste. In our research we aim to apply new laboratory techniques including high throughput microfluidic screening and machine learning to screen, understand and optimize these catalysis systems.

By miniaturizing the system and performing the experiment in a microfluidic system we are able to carefully control the flow properties and temperature gradients and tie a key characteristic of the reaction- the exotherm- to the activity of the catalyst. The critical step in the polymerization mechanism is the propagation of the polymer chain where the pi-bond in an alpha-olefin is converted to a sigma bond, releasing large amounts of energy. By monitoring this progression in a microchannel using IR thermography we are able to get a real time idea as to the concentration of monomer left in the system and thus quantify the activity of the catalyst under various conditions. This is supplemented with measurements by a Dynamic Light Scattering (DLS) instrument to quantify the chain length of the polymers, an important industrial characteristic. Overall this microfluidic platform allows us to capture relevant kinetic information in a highly controlled and repeatable fashion.

The second component which comes into play for the fast screening and optimization of these polymerization catalysts an automated reagent delivery system and control software. The reagent delivery system is constructed in such a way as to allow for a continuously variable composition of reactants- necessary for understanding broad reaction spaces. Information from the experiment is fed back to a machine learning algorithm trained on first-principles data which interprets the activity of the catalyst and updates the training. Based on the updated training the algorithm predicts a new theoretical maximum efficiency range for the catalyst and sends that particular composition to the pumping sub-system to perform the experiment. Overall this allows for the fast screening of catalysts and for the derivation of optimal operating conditions without the need for numerous manual experiments.

Overall while this methodology is currently being applied to the academic study of Zeigler-Natta catalysts, the overall goal is to enhance industrial adoption of AI and ML for the development of chemical processes. By being able to integrate highly multidimensional data and derive optimized conditions automatically we hope to enable a new path towards catalyst discovery.

Copyright © American Institute of Chemical Engineers. All rights reserved.

576554 Simultaneous Canonical Polyadic Decomposition As a Data Fusion Algorithm to Develop Pseudo-Chemistry from Spectral Data

Simultaneous Canonical Polyadic Decomposition as a Data Fusion Algorithm to Develop Pseudo-chemistry from Spectral Data

Keywords: Data Mining, Tensor Decomposition, Data Fusion, Bayesian Networks, Reaction pathway generation

In-line spectral analyzers are popularly used to obtain molecular-level information as they are fast, non-invasive, non-destructive, inexpensive and do not require sample preparation. The process data from spectral analyzers are high dimensional, multi-way, non-causal, non-full rank and have missing values. This offers a challenge in using such process data to develop causal models that hypothesize reaction pathways. This work uses the spectral datasets from Fourier Transform Infrared (FTIR) and Proton Nuclear Magnetic Resonance (^{1}HNMR) spectroscopy from the vis-breaking process of Cold Lake bitumen. The absorbance data over wavenumber/ chemical shifts are collected across different temperatures and residence time of reaction of the samples in the vis-breaker. The two spectral measurements collected across the three modes viz. temperature, residence time and absorbance/ chemical shifts represent multi-way tensorial blocks of process data. The objective of this work is to develop a data fusion framework that does not discount the tri-linear structure of the tensorial blocks of spectral measurements, accounts for the missing spectral measurements and the high-dimensionality of absorbance data across wavenumbers/ chemical shifts while aiming to incorporate complementary information about the sample from both. This is accomplished using the scheme of simultaneous canonical polyadic decomposition (CPD) where the tensorial blocks of spectral data are jointly factorized into independent factors in each mode while capturing intermodal interactions during the decomposition, resulting in a unique factorization scheme that is free from rotational and intensity ambiguities [1].

The jointly mined process data using the data fusion algorithm is then used to develop inferential models for monitoring the complex process of vis-breaking by developing pseudo-reaction networks that hypothesize chemical pathways [2]. It is to be noted that the difficulty in analytical characterization of the reacting mixture means that even the enumeration of all the major species taking part in reactions is a significant challenge for conventional methods. This is done using a probabilistic framework wherein the independent factors of a mode obtained from the simultaneous CPD are viewed as random variables with a multinomial distribution (the hyperparameters of which have a Dirichlet distribution). Bayesian networks, which are probabilistic graphical models that encode directed acyclic causal structures among the nodes of these random variables are used to develop inferential hypotheses about chemical pathways from the factors obtained from data fusion of spectral measurements. A directed path exists between nodes if it maximizes the log likelihood, which is a function of the mutual information and entropy, calculated using the probability distributions of the random variables designated as nodes (factors obtained from CPD). This amounts to using a score called the Bayesian information criterion (BIC), which is the log likelihood of the entire network (pairwise directed edges between nodes) penalized by the complexity of the network (number of edges between nodes). Heuristic greedy search score-based methods that make locally optimum choices while checking to see if directed edges between pairwise nodes maximize the penalized log likelihood are used to obtain the Bayesian networks i.e. the directed acyclic graph (DAG) encoding causal relationships among the factors obtained from simultaneous CPD. During data fusion using the simultaneous CPD, the independent factors in each mode are constrained to be non-negative so that the decomposition is physically meaningful by complying with the Beer Lambert law for spectral data, which states that the absorbance is directly proportional to the concentration of the components. Hence the independent factors in each mode from the data fusion algorithm can be physically interpreted as representing a class of chemical compounds (pseudo-component). The trilinear tensorial decomposition helps us obtain the concentration of these pseudo-components in the modes of temperature and residence time; while the third mode of wavenumbers/ chemical shifts contain the spectral signatures of the corresponding pseudo-components.

The number of independent factors in CPD is obtained using an important diagnostic called the core consistency diagnostic [3], whereby the number of factors in a CPD is obtained by fitting PARAFAC models with arbitrary factors and then casting them into Tucker models, which should produce an identity hypercube if the right number of components are used. This technique exploits the fact that tensor decompositions are higher order extensions of matrix singular value decompositions (SVD) and broadly fall into two categories: Parallel factor decomposition (PARAFAC), which represents a tensor as a sum of several rank 1 tensors, and Tucker decomposition, which is a higher order SVD. The main difference between CPD using PARAFAC and Tucker is that the number of factors is invariant across the modes in the former, making it a restricted version of the latter. When this diagnostic was used on the tensor blocks of FTIR and ^{1}HNMR data in this work, the number of factors obtained was four. The spectral signatures of the four pseudo-components obtained from the data fusion algorithm (Fig. 1) were then used to develop Bayesian networks (Fig. 2).

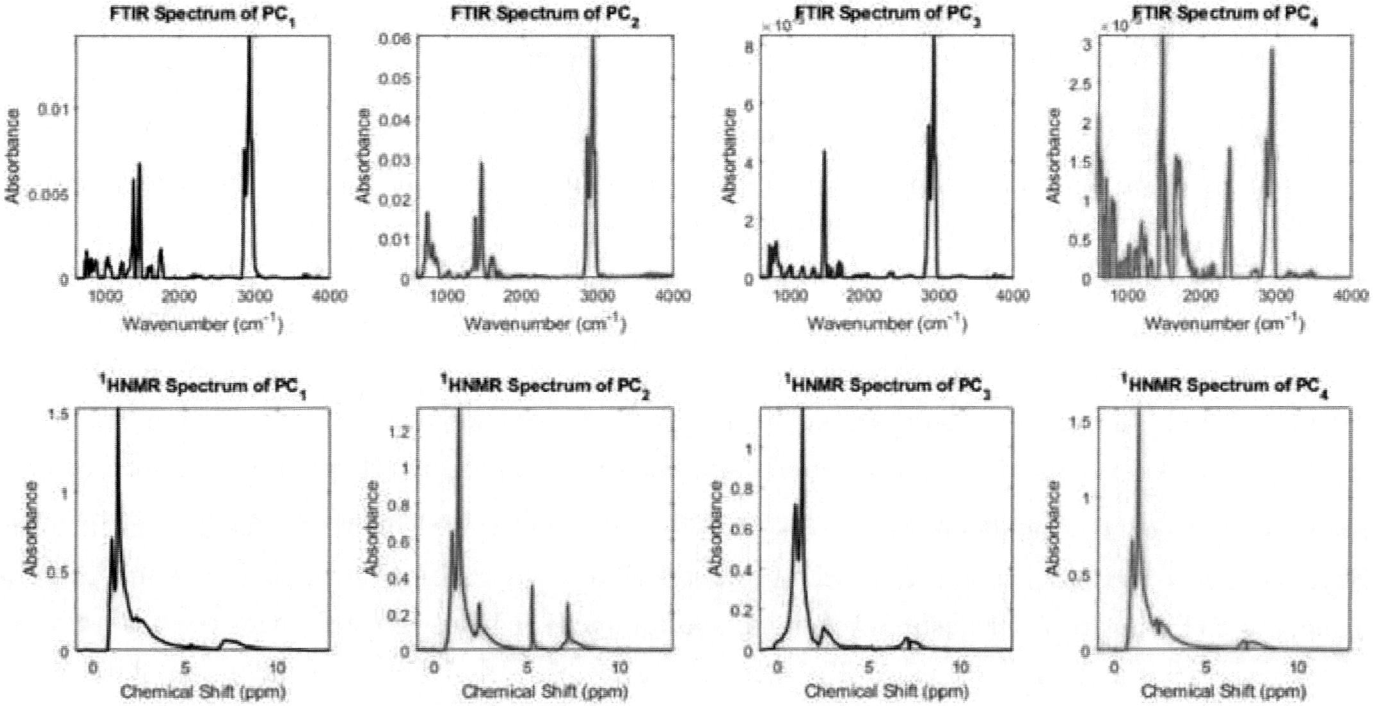

Figure 1. Spectral signatures of pseudo-components obtained from the data fusion algorithm

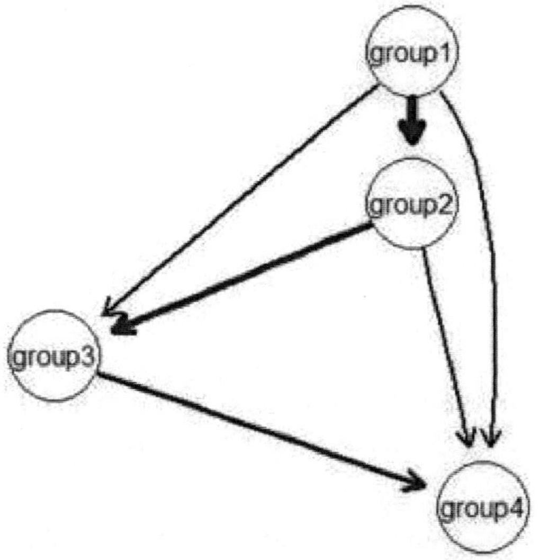

Figure 2. Bayesian networks developed from the spectral signatures of the pseudo-components (PC)

The analysis of the spectral signatures reveal the following: pseudo-component 1 (PC_1) mainly consists of carbonyl groups and cycloalkanes; pseudo-component 2 (PC_2) consists of polyaromatics, alkoxy groups, phenols, alkenes; pseudo-component 3 (PC_3) consists of aromatics, alkanes and condensed products; and pseudo-component 4 (PC_4) consists of phenols, acyls and condensed aromatics. It can therefore be hypothesized from the above Bayesian network structure that the underlying chemical reaction pathways during the vis-breaking of bitumen aim at obtaining more saturated end products through the free radical mechanism of hydrogen radical addition. However, the longer chain aliphatics crack to give more condensed polyaromatic products which are undesirable even as the end products of thermal cracking have a more aliphatic nature (alkanes and olefins).

The novelty of this work lies in the implementation of a constrained simultaneous CPD of multiple tensors using an optimization approach [4] of solving for the decision variables (factors) using the gradients of the objective function, which is formulated as the reconstruction error of the tensors from their multi-modal factors. This is an improvement over the Canonical Polyadic Alternating Least Squares (CPALS) approach, typically used for the unconstrained decomposition of just one tensor block. CPALS is not very accurate as it is not guaranteed to converge to a stationary point. Besides, the simultaneous CPD developed in this work is designed to handle missing data by imputing them by a weighting matrix in the objective function of the optimization framework. The causal DAG among the factors obtained from jointly decomposing both the tensors (FTIR and ^{1}HNMR measurements) is developed using Bayesian networks and is representative of the underlying chemical pathways among the pseudo-components. This work facilitates jointly mining spectral measurements in the framework of constrained data fusion to make the factors physically interpretable so that a first pass to building causal inferential models to generate reaction network hypotheses from process data (spectral measurements) could be implemented.

References

[1] T. G. Kolda and B. W. Bader, "Tensor Decompositions and Applications ," *SIAM Rev.*, vol. 51, no. 3, pp. 455–500, 2009.

[2] D. T. Tefera, L. M. Yañez Jaramillo, R. Ranjan, C. Li, A. De Klerk, and V. Prasad, "A bayesian learning approach to modeling pseudoreaction networks for complex reacting systems: Application to the mild visbreaking of bitumen," *Ind. Eng. Chem. Res.*, vol. 56, no. 8, pp. 1961–1970, 2017.

[3] R. Bro and H.A. Kiers, "A new efficient method for determining the number of components in PARAFAC models," *J.Chemometrics*, Vol.17,274-286, 2003.

[4] E. Acar, D. M. Dunlavy, and T. G. Kolda, "A scalable optimization approach for fitting canonical tensor decompositions ," vol. 25, no. 2, pp. 67–86, 2011.

-

-

-

Copyright © American Institute of Chemical Engineers. All rights reserved.

569295 Bridging the Synthesis-Surface Information Gap: Incorporating Spectroscopy Data into an Experimental Machine Learning Framework

One current challenge in machine learning for heterogeneous catalysis is selecting descriptors which correctly relate catalyst composition to the surface properties that directly influence activity. For machine learning to be predictive, it requires that all features input to the model are easily accessible or calculable, but collection of surface information requires measurement of synthesized catalysts. This often precludes using surface information as a predictor for catalyst activity, since the goal of a catalyst discovery framework is to predict activity without needing to synthesize new catalysts. To bridge this gap, a machine learning model is created to encode Raman spectroscopy data into spectroscopic descriptors. This model is trained to predict these descriptors from easily accessible catalyst properties, such as electronegativity or ionization energy, and experimental synthesis conditions. A similar methodology has been demonstrated for x-ray diffraction data but has yet to be applied to surface sensitive techniques such as Raman spectroscopy [1]. These intermediate surface descriptors are subsequently used alongside easily accessible catalyst properties as training data in a machine learning algorithm to predict catalyst activity for unknown catalyst formulations.

[1] A. Corma, J. M. Serra, P. Serna, M. Moliner, Integrating high-throughput characterization into combinatorial heterogeneous catalysis: Unsupervised construction of quantitative structure/property relationship models. *J. Catal.* **232**, 335–341 (2005).

Copyright © American Institute of Chemical Engineers. All rights reserved.

570431 Machine Learning and Transition Metal Chemistry: Data-Driven Comparisons of First and Second Row Complexes

The design of molecular transition metal (TM) complexes with targeted properties has great potential for homogeneous catalysis, molecular electronics and functional materials. However, the space of possible TM complexes is enormous and poorly understood, especially compared to organic chemistry where notions of chemical similarity are well established. High-throughput virtual screening of open-shell TM complexes by density functional theory can estimate the properties of novel complexes but is complicated by spin and oxidation state dependent behavior, long computation times and difficulty in reliably converging to the intended geometries and electronic states. To address these challenges, we have developed a unified strategy for the data-driven analysis and design of transition metal complexes, combining autonomous simulation, specifically engineered geometry-free representations, and fast property prediction with neural networks, uncertainty estimation and machine-learning informed workflow control. By exploiting a few thousand DFT calculations on first row TM complexes, we were able to train neural networks and use them to design new complexes with targeted spin state ordering and frontier orbital properties. By directly incorporating measures of model confidence, we are able to validate most leads at a DFT level, providing at least an order-of-magnitude increase in design efficiency compared to DFT alone. Additionally, we predict the likelihood of success of calculations to avoid performing unproductive simulations based on live analysis of the wavefunction, greatly increasing the success rate of screening and avoiding waste of computational resources. Now, we apply our automated framework to investigate properties of second-row TM complexes containing centers Mo-Rh, making a comparison to the electronic and geometric properties first-row equivalents. We assess transferability of models trained on first row data to this space and find that our outcome prediction model is transferable and can effectively reduce the number of unproductive simulations on these systems. Model transferability for property prediction is a strong function of the representation choice, and representations that directly encode periodic information can lead to improved learning both between rows and on out-of-sample errors for each row when re-training on the combined data. This demonstrates how data-driven methods can be used as a lens to examine chemical similarity and identify chemical trends in design spaces that are otherwise relatively poorly understood.

Copyright © American Institute of Chemical Engineers. All rights reserved.

570788 Active Machine Learning Model for the Discovery of Novel Material Systems and Its Application to Electrocatalysis

The design and discovery of new material systems with desired chemical and physical properties is imperative towards addressing global energy challenge. Traditional paradigms in material development are often slow, labor intensive and are not guaranteed to fully explore the expansive chemical space. Herein, we report on a robust and generalizable active learning framework for the discovery of novel and stable bulk crystal structures in the full space of crystal structural prototypes of fixed stoichiometry. Furthermore, we demonstrate the utility of this framework by the search in space of Ir-oxides, which has direct relevance to materials application as oxygen evolution reaction (OER) and oxygen reduction reaction (ORR) catalyst and support.

An active surrogate machine learning algorithm employing a Gaussian kernel was implemented to perform a rapid exploration of the Ir-oxide chemical space. The structural space was populated by all unique AB_2 and AB_3 structures from Materials Project and Open Quantum Materials Database (OQMD) databases using the prototype generator of Jain, et al available at https://www.catalysis-hub.org/prototypeSearch [1][2]). Since this approach can be easily generalized to rigorously enumerate a full structural space it also guarantees the completeness of our search. We show that the active learning framework can select the most valuable structures to perform DFT calculations on by implementing a simple acquisition criteria that biases towards low energy structures. The surrogate model resolves the most stable materials in the target space with a minimal number of DFT calculations (enough for the model's accuracy in the low formation energy region to be sufficiently low). Next, the subset of most stable IrO_2 and IrO_3 structures was computationally tested for their efficacy towards the electrochemical OER. The results for our bulk and surface calculations show the importance of highly oxided Ir-phases and provide atomistic explanation of the observed high stability and activity of the previously reported IrO_x/Sr_xIrO_3 system [3]. This work demonstrates the importance of data science approaches to accelerate catalyst discovery.

References

[1] Jain A., Bligaard T, Physical Review B **2018**, 98, 214112.

[2] Winther K., Hoffmann M., Mamun O., Boes J., Nørskov J., Bajdich M., Bligaard T., ChemRxiv **2018**,

[3] Seitz L., Dickens C., Nishio K., Hikita Y., Montoya J., Doyle A., Kirk C., Vojvodic A., Hwang, H., Nørkov J., Jaramillo T., **2016**, 353, 6303

Copyright © American Institute of Chemical Engineers. All rights reserved.

572243 A Machine Learning Framework for the Analysis and Prediction of Catalytic Activity from Experimental Data

We present a machine learning framework to explore the predictability limits of catalytic activity from experimental descriptor data that characterizes catalyst formulations and reaction conditions. Artificial neural networks are used to fuse descriptor data to predict activity and we use principal component analysis (PCA) [1] and sparse PCA [2] to project the experimental data into an information space and with this identify regions that exhibit low- and high-predictability. Our framework also incorporates a constrained-PCA optimization formulation that identifies new experimental points while filtering out regions in the experimental space due to constraints on technology, economics, and expert knowledge. This allows us to navigate the experimental space in a more targeted manner. Our framework is applied to a comprehensive water-gas shift reaction data set, which contains 2,228 experimental data points collected from the literature [3]. Neural network analysis reveals strong predictability of activity across reaction conditions (e.g., varying temperature) but also reveals important gaps in predictability across catalyst formulations (e.g., varying metal, support, and promoter). PCA analysis reveals that these gaps are due to the fact that most experiments reported in the literature lie within narrow regions in the information space. We demonstrate that our framework can systematically guide experiments and the selection of descriptors in order to improve predictability and identify new promising formulations.

[1] Jolliffe, Ian. *Principal component analysis*. Springer Berlin Heidelberg, 2011.

[2] Zou, Hui, Trevor Hastie, and Robert Tibshirani. "Sparse principal component analysis." *Journal of computational and graphical statistics* 15.2 (2006): 265-286.

[3] Odabaşı, Çağla, M. Erdem Günay, and Ramazan Yıldırım. "Knowledge extraction for water gas shift reaction over noble metal catalysts from publications in the literature between 2002 and 2012." *International Journal of Hydrogen Energy*39.11 (2014): 5733-5746.

Copyright © American Institute of Chemical Engineers. All rights reserved.

573779 Graph Convolutional Machine Learning Methods for the Predictions of Adsorption and Thermochemistry and Surface Stability

Prediction of surface phenomena such as surface stability or adsorption thermochemistry in catalysis have traditionally focused on simple coordination-based models or correlated these properties with electronic descriptors such as d-band properties. In recent work we showed that flexible coordination based fingerprints with sufficient data (thousands of calculations) were sufficient to achieve ~0.2 eV MAE accuracy for CO/H adsorption energies across a wide range of catalyst compositions. In this work we show that graph convolutional methods that learn important features and local activity from local atomic environments outperform our previous approach and are capable of achieving ~0.14 eV accuracy across thousands of surfaces with >30 unique elements while simultaneously predicting the contribution of each surface atom to the overall property. We also demonstrate this approach for a new dataset of intermetallic surface energies and show that these methods are capable of qualitatively predicting dominant nanoparticle facets. This works for sufficiently large datasets and demonstrates the transition from small-data (featurization and machine learning) to large data (deep learning) approaches for the prediction of surface phenomena in catalysis.

Copyright © American Institute of Chemical Engineers. All rights reserved.

570712 Influence of Interfaces on the Structural Evolution in Ionic Liquid-Ionic Liquid Mixtures

The flexibility to manipulate various interactions in ionic liquids by changing the cation, anion, or the various functional groups on the ions has drawn researchers to design tailor-made ionic liquids for a given application. Another approach that is gaining importance in tuning ionic liquid properties is to form binary, ternary, and reciprocal ionic liquid mixtures. The research in our group has demonstrated that the binary ionic liquid mixtures in which two anions are combined in different molar ratios for a given cation can lead to either native or non-native structures in the bulk depending on the difference between the hydrogen bonding ability of the participating anions. The non-native structures differ markedly from those for the pure ionic liquid analogues and impact the physical dissolution mechanism for gases as CO_2. Given the importance of ionic liquids as electrolytes in the next-generation of batteries, it is likely that such ionic liquid mixtures will increasingly be investigated for their ability to transport ions to and from charged surfaces. However, very little is currently known regarding how the presence of a surface or an interface affects the local and long-range structures of binary ionic liquids. In this talk, we will present our initial findings in this field by examining the response of binary ionic liquid mixtures comprised of a single cation in combination with two anions when presented with a graphite surface. We will compare and contrast these structures to those obtained when binary ionic liquid mixtures are formed with a single anion and two cations and a graphite surface is introduced.

Copyright © American Institute of Chemical Engineers. All rights reserved.

560557 Interfacial Structure of Ionic Liquids Studied with Coarse-Grained Molecular Dynamics

The increased attention and utilization that ionic liquids have gained in the last couple of decades has been fueled not only by realization of their useful and unique physical and chemical properties, but also by a relatively new perspective in their analysis of recent years. All-atom molecular dynamics (MD) simulations, which expose these fluids' customizability for use in vast technical applications, have adapted to produce coarse-grained (CG) simulation models for more computationally efficient computation of various systems. By reducing the mechanical degrees of freedom, these CG models make possible the simulation of larger time and length scales, enabling observation of complex structural behavior present in systems such as those of ionic liquids. However, even well-constructed CG models will not necessarily produce correct dynamic behavior, because reduction in degrees of freedom reduces "effective friction" between CG beads. To address this issue, the more recently developed probability distribution function coarse-graining (PDF-CG) method has been implemented. In this method, effective friction is reintroduced into the CG system through the stochastic Langevin equation. With the proposed approach, accurate prediction of structure and dynamics for CG systems like ionic liquids has been observed. This model can provide a key component in understanding ionic liquid behavior at various interfaces and chemical compositions, bridging the gap between the fluids' impressive properties and development of next generation technology. The present work details application of the PDF-CG method to simulate two sets of ionic liquids: (a) *N*-methyl-*N*-butylpyrrolidinium bis(trifluoromethylsulfonyl)-imide ([pyr14][TFSI]), and (b) 1-ethyl-3-methylimidazolium boron tetrafluoride ([EMIM][BF_4]). First outlined is the development and validation of CG models, in which CG force fields were successfully obtained to produce correct structure and dynamics of the CG system in reference to all-atom MD simulations for both liquids. The system size was then increased by one order of magnitude. The developed CG models for both sets of ionic liquids were used to perform CG-MD simulations at liquid-vacuum interfaces to study inhomogeneous interfacial structure and density profiles that have been experimentally observed.

Copyright © American Institute of Chemical Engineers. All rights reserved.

568442 Spatially Varying Chemical and Physical Properties Modulate Interfacial Hydrophobicity of Self-Assembled Monolayers

Water, while simple in structure, contains many unusual physicochemical properties such as a distinct polar charge distribution which leads to the formation of hydrogen bonds. In solutions, these strong water-water interactions lead to relatively weak or unfavorable interactions with nonpolar solutes which indirectly drives the association of hydrophobic materials. This indirect driving force is known as a hydrophobic interaction (HI) which is critical to protein folding/misfolding, ligand-substrate binding, and colloidal self-assembly. Understanding the relationship between HIs and the properties of functionalized surfaces is vital to the design of new materials that enhance control over complex self-assembly and adsorption in aqueous environments. Conventionally, the strength of HIs between surfaces is attributed to interfacial chemical properties, such as the available nonpolar solvent-accessible surface area. However, recent experiments demonstrated that HIs between functionalized self-assembled monolayers (SAMs) with similar areas depended on specific chemical and physical properties, such as functional group chemistry, surface patterning, and SAM order (crystallinity).

Experiments using an atomic force microscope (AFM) revealed specific chemical and physical properties have unexpected effects on the strength of HIs. In this work, we used molecular dynamics simulations to investigate with atomistic resolution the origin of those effects. We found that the hydrophobicity of SAMs can be altered by changing interfacial properties (*e.g.*, alkyl chain length, functional group chemistry, and ligand composition). Specifically, we observed functional group chemistry modulates the hydrophobicity of the adjacent nonpolar domains in mixed SAMs. Further, we investigated the roles of molecular fluctuations and SAM order to determine the key component influencing nonpolar SAM hydrophobicity. We find that spatially varying SAM order leads to interfaces that appear chemically heterogeneous. We identified that perturbations to the nanoscale structure of nearby water molecules can explain the effects of interfacial properties on hydrophobicity. These perturbations were then quantified to provide a mechanism to explain how spatially varying interfacial properties modulate hydrophobicity. These physical insights can then be applied to new experimental systems to predict interfacial hydrophobicity.

Copyright © American Institute of Chemical Engineers. All rights reserved.

577158 The Protein Surface Hydrophilic Lipophilic Balance As a New Approach for Characterizing Large Biosurfactants

Surfactants are molecules with the capacity to decrease the interfacial tension between two immiscible phases. Within the non-ethoxylated surfactants, it is possible to find an emerging group of biomolecules derived from microorganisms; these are called biosurfactants or bioemulsifiers. As a case study, there is a growing tendency in the food and cosmetic products industry to replace surfactants derived from petroleum or chemical synthesis, with "green" alternatives such as biosurfactants. The latter are formed from amino acids, fatty acids, carboxylic acids or sugars with low molecular weight (e.g. glycolipids), or high molecular weight (e.g. proteins, lipoproteins, heteropolysaccharides and lipopolysaccharides). Some of them are proteins with a large amphiphilic structure that have more than one hydrophobic zone. Depending on their molecular weight, biosurfactant have the capacity to emulsify and lead to interfacial tension reduction. [1].

Over the years, the hydrophilic – lipophilic balance (HLB) metric has been employed to characterize their behavior, and technological capacity to form stable emulsions with a determinate lipophilic phase. This approach was originally designed for ethoxylated surfactants, which vary widely from other types of surfactants from the molecular point of view. Polyethoxilated surfactants exhibit a non-conventional interaction pattern with water molecules; for example, there is a significant conformational change associated with the temperature during the emulsification process. For this reason, some researchers have suggested that it is not possible to extrapolate the HLB parameter to cationic and ionic surfactants. For their part, Yamashita and Sakamoto [2], [3] have argued that the HLB is not a good descriptor for non-POE surfactants. In spite of all those findings, researchers and practitioners still use the HLB to describe surfactant emulsification capacity. In this context, the HLB value continues to be a universal tool for comparing surfactants [4], [5]. To address the inadequacy of the HLB metric, over the years, several HLB-like concepts have been proposed [2], [6], [7]. Despite the fact a lot of research has been done to understand monomeric surfactants behavior, the characterization of polymeric biosurfactants, especially proteins, are conspicuously absent.

Unfortunately most of the few proposed models for biosurfactants have been limited to monomeric biosurfactants (e.g. rhamnolipids [8]–[10]). As a result, little is known about how proteins can decrease the interfacial tension between immiscible mixtures [11], [12], and even less is known about how to formulate products using it. Since the relevant surfactants are smaller molecules than proteins, their three-dimensional structure does not have a significant effect; thus, proteins do not factor into the assessment. We note that due to the big structure of a protein, it makes sense that those amino-acids located deep inside the protein molecule well away from the surface, will not interact with water or oil molecules. Instead, the amino-acids located on the protein surface have the most effect on the protein emulsification capacity [13].

In recent years, membrane proteins (such as OmpA, a membrane protein of mainly Gram-negative bacteria) have become an active area of research, not only for their natural role in biological processes but also for their potential impact on industrial applications. Most of the interest centers on their amphiphilic nature; specifically with regard to OmpA, the capacity to stabilize emulsions of dodecane in water[11] has exciting and widespread implications. Those proteins have a large amphiphilic structure and more than one hydrophobic zone.

We have focused on the surface activity of two membrane proteins of *Escherichia coli,* (OmpA and OmpN) in order to compare their emulsifier capacity within a media composed of water and triglycerides. Using the Kawakami equation [14], the HLB of a surfactant can be determined through the relation between the molecular weight of its hydrophilic groups and its lipophilic groups. Following up this idea, and considering *Ala, Cys, Phe, Ile, Leu, Met*, and *Val*, as the hydrophobic residues of a protein, we applied the Kawakami equation to the entire protein sequence which results in an HLB of 11.2 and 11.6 for OmpA and OmpN respectively. Although, experimental results showed that OmpA and OmpN have different behaviors when they are used as emulsifier agents in o/w emulsions made with vegetable oils. To resolve this, we performed molecular dynamic simulation of OmpA and OmpN in a water – triglycerides media. Changes in solvent accessible surface area (SASA) for both proteins were detected, and they were considered an indicator of protein flexibility. Additionally, we found when we calculate the HLB value taking into account only the residues located at the outer layer (7 Å) surface of each protein (during the final stage of the production of the MD simulation), we obtained a HLB value of 9 to 10 for OmpA, while of 6 to 7 for OmpN. We call this new HLB value at the surface approach the "protein surface hydrophilic lipophilic balance (PS – HLB)". This calculation allows us to consider only the surface residues as responsible of the interaction with water and hydrophobic molecules at the interface. The PS-HLB values are more accurate than the traditional HLB for describing the experimental behavior of those two proteins when emulsifying vegetable oils.

References

[1] W. Zhang, K. K. S. Randhawa, P. K. S. M. Rahman, C. Uzoigwe, J. G. Burgess, and C. J. Ennis, "Bioemulsifiers are not biosurfactants and require different screening approaches The Physiological Roles of Biosurfactants and Bioemulsifers are Based on their Physicochemical Properties," vol. 6, 2015.

[2] Y. Yamashita and K. Sakamoto, "Hydrophilic-Lipophilic Balance (HLB): Classical Indexation and Novel Indexation of Surfactant," in *Encyclopedia of Biocolloid and Biointerface Science 2V Set*, Hoboken, NJ, USA: John Wiley & Sons, Inc., 2016, pp. 570–574.

[3] S. Abbott, *Surfactant Science : Principles and Practice*. 2015.

[4] S. K. Satpute, A. G. Banpurkar, P. K. Dhakephalkar, B. MI.M., and B. A. Chopade, "Methods for investigating surfactants and bioemulsifiers: a review," *Crit. Rev. Microbiol.*, vol. 30, no. 2, pp. 127–144, 2010.

[5] ICI Americas Inc and I. Americas, *The HLB system a time-saving guide to emulsifier selection*. 1980.

[6] Y. Nakama, "Surfactants," in *Cosmetic Science and Technology: Theoretical Principles and Applications*, 2017, pp. 231–244.

[7] M. Álvarez Vanegas, A. Macías Lozano, V. Núñez Vélez, N. Garcés Ferreira, H. Castro Barrera, O. Álvarez Solano, and A. Fernando González Barrios, "Molecular dynamics approach to investigate the coupling of the hydrophilic–lipophilic balance with the configuration distribution function in biosurfactant-based emulsions," *J Mol Model*, vol. 19, pp. 5539–5543, 2013.

[8] T. T. Nguyen and D. A. Sabatini, "Formulating alcohol-free microemulsions using rhamnolipid biosurfactant and rhamnolipid mixtures," *J. Surfactants Deterg.*, vol. 12, no. 2, pp. 109–115, 2009.

[9] K. K. Sekhon Randhawa and P. K. S. M. Rahman, "Rhamnolipid biosurfactants—past, present, and future scenario of global market," *Front. Microbiol.*, vol. 5, p. 454, 2014.

[10] D. E. Otzen, "Biosurfactants and surfactants interacting with membranes and proteins: Same but different? ," *BBA - Biomembr.*, vol. 1859, pp. 639–649, 2016.

[11] S. M. Aguilera-Segura, A. P. Macías, D. Carrero-Pinto, W. L. Vargas, M. J. Vives-Florez, H. E. Castro-Barrera, O. A. Álvarez, and A. F. González-Barrios, "Escherichia coli's OmpA as Biosurfactant for Cosmetic Industry: Stability Analysis and Experimental Validation Based on Molecular Simulations," in *Advances in Computational Biology SE - 38*, vol. 232, L. F. Castillo, M. Cristancho, G. Isaza, A. Pinzón, and J. M. C. Rodríguez, Eds. Springer International Publishing, 2014, pp. 265–271.

[12] S. M. Aguilera-Segura, V. Núñez-Vélez, L. Achenie, O. Álvarez-Solano, R. Torres, and A. F. González-Barrios, "Peptides design based on transmembrane Escherichia coli's OmpA protein through molecular dynamics simulations in water–dodecane interfaces," *J. Mol. Graph. Model.*, vol. 68, pp. 216–223, 2016.

[13] S. Nakai, E. Li-chan, and E. Li-Chan, "Critical Reviews in Food Science and Nutrition Recent

advances in structure and function of food proteins: QSAR approach Recent Advances in Structure and Function of Food Proteins: QSAR Approach," *Crit. Rev. Food Sci. Nutr.*, vol. 33, no. 6, pp. 477–499, 1993.

[14] Y. Yamashita, R. Miyahara, and K. Sakamoto, "Emulsion and emulsification technology," in *Cosmetic Science and Technology: Theoretical Principles and Applications*, 2017, pp. 489–506.

[15] A. V Finkelstein and O. B. B. T. Ptitsyn, *Protein Physics*. Amsterdam: Academic Press, 2016.

[16] C. N. Pace, S. Treviño, E. Prabhakaran, and J. M. Scholtz, "Protein structure, stability and solubility in water and other solvents.," *Philos. Trans. R. Soc. Lond. B. Biol. Sci.*, vol. 359, no. 1448, pp. 1225-34; discussion 1234–5, Aug. 2004.

Copyright © American Institute of Chemical Engineers. All rights reserved.

565492 Conformational and Dynamical Behaviors of Polyelectrolytes Under Spatial and Dielectric Confinement

Understanding the conformational and dynamic behaviors of polyelectrolytes inside viral capsids and lipid vesicles is crucial for gene delivery and drug loading applications. In this talk, we investigate the collective effects of dielectric mismatch and spatial confinement on the equilibrium behaviors of a coarse-grained model polyelectrolyte with explicit counterions confined in a spherical cavity. We show that surface polarization due to dielectric mismatch between the media inside and outside the cavity strongly influences the equilibrium conformations of the polyelectrolyte. We will also characterize the dynamical behavior of the confined polyelectrolyte at different length scales by analyzing the relaxation times of the Rouse modes in reference to those of the unperturbed counterpart. The roles of confinement volume, surface charge density, electrostatic strength between the monomers and multivalent counterions will be discussed. The findings therein offer insight into the effects of dielectric mismatch, which were often overlooked in many studies on confined polyelectrolytes.

Copyright © American Institute of Chemical Engineers. All rights reserved.

573860 Gibbs Ensemble Monte Carlo Simulations for Additive Loading in Surfactant Bilayers

Understanding solute uptake into soft micro-structured materials, such as bilayers and micelles, is of specific interest in the pharmaceutical, agricultural, and personal care industries. Currently, there are continuum-solvation methods available that can rapidly estimate the amount of a solute that will load into these soft micro-structured materials. However, these methods do not give any insight into the structural orientation and positioning of the solutes at a molecular level. To obtain this insight, we utilize the Shinoda-Devane-Klein (SDK) force field in conjunction with configurational-bias Monte Carlo simulations in the osmotic Gibbs ensemble. One simulation box contains a lamellar system comprised of a bilayer formed by triethylene glycol mono-*n*-decyl ether (C10E3) surfactants surrounded by water with a 50/50 surfactant/water weight ratio. This simulation cell is allowed to undergo volume and shape fluctuation, by changing either the z cell length independently or the x and y cell lengths in a coupled fashion and is found to have a stable lamellar phase in agreement with experiment. The other simulation box is a vapor-phase reservoir containing solute molecules at a constant applied pressure, and only solute molecules are allowed to transfer between the two simulation boxes. We study both the adsorption isotherm and structural effects of *n*-nonane, 1-hexanol, and ethyl butyrate on the bilayer at several different reduced pressures with respect to the saturated vapor pressure of the solutes. Additionally, we study the binary adsorption of *n*-nonane and 1-hexanol and observe a possible synergistic enhancement of their uptake.

Copyright © American Institute of Chemical Engineers. All rights reserved.

564046 Understanding the Role of Cellulose Nanocrystals As Novel Shale Hydration Inhibitors for Water-Based Drilling Fluids

During oil well drilling operations, water-based drilling fluids (WDFs) are used to lubricate and cool down equipment, carry rock cuttings to the surface, and prevent excessive uptake of water by the clay minerals in the shale formation. Shale hydration issues can lead to reservoir instability problems and wellbore damage, which are responsible for important economic losses (>$500M/year) during drilling operations. Therefore, the development of effective and environmentally friendly hydration inhibitors for WDFs is a very important issue in the petroleum field [1,2]. Recent studies by our collaborator Qinglin Wu (LSU) suggests that biodegradable cellulose nanocrystals (CNCs) can be used as additives to WDFs to inhibit clay hydration and enhance WDF filtration and rheology [3-5]; however, the physics behind this observations remains uncertain. As an initial step, here we probe the interactions between CNCs and clays (Wyoming-type montmorillonites with potassium ions) in the presence of water, using all-atom classical molecular dynamics simulations. We will present and discuss our results for the potential of mean force as a function of the distance between different facets of the CNCs and the montmorillonites.

1. Anderson, R. L.; Ratcliffe, I.; Greenwell, H. C.; Williams, P. A.; Cliffe, S.; Coveney, P. V. Clay swelling — A challenge in the oilfield. Earth-Science Reviews 2010, 98, 201-216.
2. Suter, J. L.; Coveney, P. V.; Anderson, R. L.; Greenwell, H. C.; Cliffe, S. Rule based design of clay-swelling inhibitors. Energy & Environmental Science 2011, 4, 4572-4586.
3. Li, M.-C.; Wu, Q.; Song, K.; De Hoop, C. F.; Lee, S.; Qing, Y.; Wu, Y. Cellulose Nanocrystals and Polyanionic Cellulose as Additives in Bentonite Water-Based Drilling Fluids: Rheological Modeling and Filtration Mechanisms. Industrial & Engineering Chemistry Research 2016, 55, 133-143.
4. Song, K.; Wu, Q.; Li, M.-C.; Wojtanowicz, A. K.; Dong, L.; Zhang, X.; Ren, S.; Lei, T. Performance of low solid bentonite drilling fluids modified by cellulose nanoparticles. Journal of Natural Gas Science and Engineering 2016, 34, 1403-1411.
5. Song, K.; Wu, Q.; Li, M.; Ren, S.; Dong, L.; Zhang, X.; Lei, T.; Kojima, Y. Water-based bentonite drilling fluids modified by novel biopolymer for minimizing fluid loss and formation damage. Colloids and Surfaces A: Physicochemical and Engineering Aspects 2016, 507, 58-66.

Copyright © American Institute of Chemical Engineers. All rights reserved.

572961 Molecular-Level Characterisation of Crystal/Solution Interfaces: The Case of Ibuprofen

In the pharmaceutical industry active pharmaceutical ingredients (APIs) are often manufactured as crystalline solids. The solid form of a drug provides several advantages in the isolation, handling and storage of the material, such as higher purity, improved flow, and higher chemical and physical stability. Probably the most critical decision in the process and product design of the API manufacture is the choice of the solvent for the crystallisation step. The solvent can affect the size and shape distribution of the resulting crystalline particles, which in turn can have a significant impact on its chemical and physical properties, as well as it bioperformace [1]. Predicting crystal shapes is therefore desirable in order rationally identify favourable crystallization conditions and develop efficient API manufacturing processes.

The morphology of solution-grown crystals is affected by the solvent due to specific surface-solvent interactions. These interactions can alter the relative growth rate of morphologically dominant crystallographic faces, resulting in a modified crystal shape. Mechanistic models and molecular simulations have been used over the last decade in attempt to account for solvent effects and provide an accurate crystal shape prediction. Developing such models is challenging due to the diversity of factors that may affect growth rates, usually present within a single system. Factors such as solubility, formation of hydrogen bonds, strong non-bonded interactions, degree of conformational flexibility of both solute and solvent molecules can all play a role in determining the growth mechanism at the crystal surface or have an impact on the rate-determining step of the process. The level of complexity increases further since crystallographic faces within the same system expose different functional groups to the solution and a different rate-altering factor dominates. Solvent-specific crystal shape prediction has been achieved for several systems [2,3], however, the development of a universal and systematic crystal shape prediction model still presents outstanding challenges [4].

In our research we lay the groundwork of tackling these challenges and propose a systematic way of accounting for the effects of specific surface-solvent interactions on the morphology of solution-grown crystals. Our efforts are focused in two main directions. Firstly, we study the role of conformational flexibility of the solute molecule in the context of crystal growth [7]. To this aim, we study the conformational rearrangement of ibuprofen with the aid of molecular simulations. We investigate the thermodynamics and dynamics of ibuprofen conformational transitions in the states involved in the process of incorporation of a molecule in the crystal bulk (in solution, at the surface/solution interface and in the crystal lattice). Secondly, we investigate the solvent behaviour as a function of distance from the morphologically dominant crystal faces of ibuprofen for several solvents. We obtain information on the thermodynamics and dynamics of solvent molecules at the crystal-solution interface. We account for the free energy difference between a molecule adsorbed on the crystal surface compared to bulk solution, as well as the exchange rate of an adsorbed solvent molecule with a molecule from the bulk solution or a subsequent solvent layer. In our study we consider the morphologically dominant {002}, {011}, {110}, {100} polar and {100} apolar ibuprofen crystal faces in 9 different solvents – water, 1-butanol, toluene, cyclohexanone, cyclohexane, acetonitrile, trichloromethane, methanol and ethyl acetate.

This work is carried out with the aid of molecular dynamics (MD) simulations. We investigate the conformational flexibility of ibuprofen by applying state of the art enhanced sampling molecular simulation techniques [5]. State-to-state kinetics and conformational rearrangement mechanisms are then reconstructed within the theoretical framework of Markov State Models [6]. Studying the structure of the solution in the presence of a crystal surface is achieved by extensive post-processing of MD trajectories with Plumed 2.3[8].

Our study reveals that the conformational free energy landscape of ibuprofen is strongly dependent on the environment and the relative position with respect to the crystal surface. While in solution we find a total of 6 conformers, all characterised by non-negligible populations, in the crystal bulk the distribution of conformers consists almost exclusively of one conformer. At the crystal/solution interface the distribution of conformers varies depending on the solvent and on the structure of the interface. The system relaxation time is found to also depend on both the solvent and the environment. Furthermore, when moving from solution to the crystal surface the mechanism of conformational rearrangement changes significantly. Last but not least, the structure of the solid/liquid interface is found to be strongly dependent on the solvent, affecting the conformational flexibility of molecules at the surface.

The investigation of the solution structure reveals that the arrangement of solvent molecules in the solution and the stability of the absorbed state varies greatly depending on their interaction with the morphologically dominant crystal faces. We find that while for some surface/solvent combinations an adsorbed state is practically non-existent, in other cases surface-solvent interactions promote a stable adsorbed state and can furthermore induce structuring effect in the solution. In particular long-range effects are observed in the case of surface-induced stacking of solvent molecules, which propagates beyond the solvent molecules in direct contact with the crystal surface. In such cases the rate of exchange of solvent molecules between adsorbed layers is remarkably slow, with characteristic exchange times up to three orders of magnitude slower than in cases in which long-range layering is not observed.

In the context of crystal growth our findings have important consequences. On the one hand, we find that incorporation of a solute molecule from bulk solution to an adsorbed state may be limited by conformational rearrangement due to the difference in conformational distribution between states. This process, however can also be significantly impaired by the structure of the solution. In fact, for cases where we observe solvent layering, the kinetics of a mass transport driven process such as rough growth can be significantly affected. Furthermore, under the assumption that a lifetime of an adsorbed state of a solvent molecule is correlated with the vacation of a kink site, the solvent kinetics at the surface/solution interface may carry a great significance in the growth rate of the particular face and provide a crude approximation of the solvent effect on the crystal shape. We find that the surface-solvent interactions also affect the degree of flexibility of a solute molecule in the crystal surface and the resulting freedom of conformational rearrangement can affect the rate of solute ordering at the crystal surface. Overall, we believe our work paves the way towards a formulation of solvent-dependent kinetic models for the growth and dissolution of flexible molecules.

References

1. N. Variankaval, A. Cote, *AIChe Journal*, **2008**, *54(7),* 1682
2. M. Salvalaglio, T. Vetter, M. Mazzotti, M. Parinello, *Angewandte Chemie Int Edition*, **2013**, *52(50)*, 13369
3. J. Li, M. Doherty, *Crystal Growth and Design,* **2017**, *17(2),* 659
4. E. Elts, M. Greiner, H. Briesen, *Crystals,* **2017**, *7*, 288
5. A. Barducci, G. Bussi, M. Parrinello, Phys, Rev. Letters, 2008, *100*, 020603
6. G. R. Bowman, V. S. Pande, An introduction to Markov State Models and Their Apllication to Long Timescale Molecular Simulations, Springer, USA
7. V. Marinova, G.P. Wood, I. Marziano, M. Salvalaglio *JCTC*, **2018**, *14(12),* 6484
8. G. Tribello, M. Bonomi, D. Branduardi, C. Camilloni, G. Bussi, *Computer Physics Communications*, **2013** *185*, 604

Copyright © American Institute of Chemical Engineers. All rights reserved.

564365 Molecular Simulation-Guided Design of Hydrophilic-Hydrophobic Interfaces for Hybrid MOF Microtanks for Methane Storage

Given the abundance of natural gas in the U.S., the widespread use of methane-powered vehicles seems an enticing prospect. The incumbent method to store methane in on-board tanks is by densification to 265 cc(STP)/cc through compression to 250 bar. However, the high storage pressure creates safety concerns that require the use of cumbersome tanks made of expensive materials to be addressed. Adsorption-based methane storage, which exploit gas-solid interactions to densify methane within a porous adsorbent has been considered one of the most promising ways to reduce the storage pressure. However, a series of high throughput screening studies have indicated that there may be material design constraint that fundamentally prevent reaching desirable methane adsorption levels in "stand alone" adsorbents such as metal-organic frameworks (MOFs).

A "hybrid" material design with potential to overcome those limits was introduced in experiments by Carreon and coworkers [1]. In the hybrid design, a millimeter-sized pellet of a porous material is coated by a thin membrane of the second porous material. The pellet can be filled with methane at high-pressure and held in place by reversibly adsorbing "sealant" molecules onto the thin membrane, blocking the stored methane from going out and keeping the external pressure low. While the operating principle was demonstrated experimentally, the described hybrid design is far from being optimized. Here, we first present grand canonical Monte Carlo simulations in model pores (graphene and graphene oxide) to demonstrate that a design principle that should be pursued in these hybrid systems is to make the pellet material hydrophobic and the membrane material hydrophilic, which creates a desirable distribution of methane and a hydrophilic sealant between the pellet and the membrane to prevent methane escape, as well as sealant invasion of the pellet. Second, we present a computational high throughput screening of a 2,000+ database of computational MOF prototypes constructed using the Topologically-Based Crystal Constructor (ToBaCCo) code to elucidate how to tune the chemistry and structure of MOFs to attain the desirable hydrophobic and hydrophilic character of the pellet and membrane materials for this methane storage application.

[1] Song et al. Nano Lett., 2016, 16, pp 3309–3313

Copyright © American Institute of Chemical Engineers. All rights reserved.

571027 Data-Driven Design of Self-Assembling Pi-Conjugated Oligopeptides (Invited Talk)

Synthetic oligopeptides containing pi-conjugated cores present an attractive molecular building block for the self-assembly of biocompatible nanoaggregates with engineered optical and electronic properties. The peptidic and pi-conjugated blocks can be tailored to control the aggregation morphology and emergent optoelectronic properties such as electronic delocalization, electron/hole transport, and optical absorption spectrum. The design space of possible peptide sequences and pi-conjugated cores is so large as to make Edisonian trial-and-improvement of the structure and properties of the self-assembled aggregates intractable by either experimentation or simulation. This motivates a data-driven guided search wherein surrogate models trained over data for a small number of oligopeptide chemistries informs an active learning protocol to identify the next most promising chemistries to study. In this work we combine coarse-grained molecular dynamics (CGMD) simulations, variational autoencoders (VAEs), and Gaussian process regression (GPR) to establish an integrated pipeline for the rational exploration of pi-conjugated oligopeptide chemical space. CGMD is used to predict the quality of core-core alignment within the self-assembled nanoaggregates as a measure of fitness, VAEs are trained from scratch to identify a low-dimensional parameterization of chemical space within which to perform the search, and GPR with active learning is used to efficiently navigate the space through an acquisition function balancing exploitation (i.e., chemistries predicted to possess good assembly behavior) and exploration (i.e., searching in under-sampled regions of chemical space). This protocol leads to the identification of new unanticipated pi-conjugated oligopeptide chemistries for all-atom molecular modeling, electronic structure calculations, and experimental synthesis and characterization.

Copyright © American Institute of Chemical Engineers. All rights reserved.

576972 Data-Driven Design of Functional Amyloid Biomaterials

The design of novel amyloid biomaterials with diverse structuresand functions is an upcoming field with future tremendous applications in biomedicine and biotechnology. However, the design of functional amyloid biomaterials with the capacity to bind to ions and compounds, or amyloid biomaterials with cell-attachment properties has been significantly limited until currently experiments primarily on researchers' intuition on which mutations should be added to amyloids to transform them into functional biomaterials. During the last years, we have been developing the first computational protocol with the ability to transform amyloid peptide scaffolds into functional amyloid biomaterials. According to the protocol, suitable amyloid short peptide structures with non-ß-sheet residue positions at the termini (referred to as amyloid designable scaffolds [1]) are transformed into functional amyloid biomaterials, by computationally designing these positions to bear a specific functionality. Specifically, the protocol searches in an optimization-based approach for combinations of modifications, which according to data analysis, can yield a desired functionality and which can produce energetically favorable solutions.

The talk will focus on the design of functional amyloid biomaterials binding to cesium ions, at which the designed amino acids were introduced to the amyloid designable scaffold to mimic how amino acids bind to cesium ions according to data analysis on experimentally resolved structures binding cesium ions [2]. As part of the protocol the optimum designs were computationally validated using a series of simulations and structural analysis to select the top designed peptides which were experimentally validated [2]. The talk will also present recent applications of the computational protocol in additional studies, emphasizing on how data-driven analysis can lead to the formation of functional amyloid biomaterials.

[1] Kokotidou, C.; Jonnalagadda, S. V. R.; Orr, A. A.; Seoane-Blanco, M.; Apostolidou, C. P.; van Raaij, M. J.; Kotzabasaki, M.; Chatzoudis, A.; Jakubowski, J. M.; Mossou, E.; Forsyth, V. T.; Mitchell, E. P.; Bowler, M. W.; Llamas-Saiz, A. L.; Tamamis, P.; Mitraki, A., A novel amyloid designable scaffold and potential inhibitor inspired by GAIIG of amyloid beta and the HIV-1 V3 loop. FEBS Letters 2018, 592 (11), 1777-1788.

[2] Jonnalagadda, S. V. R.; Kokotidou, C.; Orr, A. A.; Fotopoulou, E.; Henderson, K. J.; Choi, C.-H.; Lim, W. T.; Choi, S. J.; Jeong, H.-K.; Mitraki, A.; Tamamis, P., Computational Design of Functional Amyloid Materials with Cesium Binding, Deposition, and Capture Properties. The Journal of Physical Chemistry B 2018, 122 (30), 7555-7568.

Copyright © American Institute of Chemical Engineers. All rights reserved.

573714 Exploration of Interresidual Contacts in Combined Coarse Graining and Experiment Directed Simulations of Aβ42

Experiment Directed Simulations (EDS) and coarse-grained molecular dynamics (MD) were combined to improve the simulation of Aβ42 peptide fragments and to match their computed chemical shifts to experimental results. EDS is a maximum entropy biasing technique that minimally changes (biases) the potential energy of a simulation so that some constraint is satisfied. Constraints are usually an agreement between average position dependent variables from experiments and simulations. EDS has been found to indirectly improve dynamical and structural properties in different systems like peptides, water models, proteins, and lithium ions. EDS has been used in ab-initio, atomistic MD, and coarse-grained MD simulations. Coarse-grained EDS simulations improved the accuracy of actin monomer simulations by reproducing the fluctuations present in the larger actin filaments. In this work, EDS and coarse-graining are used to understand intramolecular and intermolecular interactions in Aβ42. Aβ42 is a 42-residue long amyloid peptide that is associated with Alzheimer's disease. Atomistic MD simulations of multiple chains of 21-30 fragment of amyloid peptide with EDS resulted in their self-assembly previously. Knowing the process of self-assembly in whole Aβ42, including the dynamics and governing interactions will help find therapeutics that halt these processes. In order to study these interactions in Aβ42, we divided Aβ42 into 5 fragments (each fragment consists of 10 residues) with overlapping ends and simulated all of the possible combinations of these 5 fragments in separate simulations. We used chemical shifts from nuclear magnetic resonance spectroscopy results to bias the fragments with chemical shifts from the whole Aβ42 peptide in EDS. All-atom MD simulations were performed with EDS and enhanced sampling in GROMACS simulation engine. We used force matching bottom-up approach to map all-atom forces to a mapped coarse-grained potential. We classified which fragments and which residues interact the most with each other. When EDS bias had been applied, we saw behavior that was predicted in the past experimental literature. During coarse-graining, the interactions were observed between specific fragments of Aβ42 and not with others, which could lead to self-assembly in the full Aβ42.

Copyright © American Institute of Chemical Engineers. All rights reserved.

566350 Conjugation of DOPA and Amyloid Structures for Molecular Design of Underwater Adhesives

Adhesives capable of sticking to various types of surfaces in underwater and high moisture conditions are required for a large array of applications such as marine coatings, sealants, medical devices, and wet living tissue repair. However, typical industrial adhesives are primarily developed for dry applications and perform poorly in wet environments or are highly specific to particular surfaces and not suitable for broad use. A viable and cost-effective strategy to address this issue is to develop generic adhesives that can be used for multiple types of surfaces in water or high moisture conditions.

One approach researchers have taken to develop new adhesives is to draw inspiration from the natural glues produced by aquatic organisms capable of strong, moisture-resistant adhesion to various surfaces. Analysis of these glues has shown that these organisms' abilities to adhere to a multitude of surfaces involve L-3,4-dihydroxyphenylalanine (DOPA) and functional amyloid nanostructures.

The objective of this project is to computationally design bio-inspired underwater adhesives that are stronger than those currently reported in the literature by analyzing and combining the adhesive capabilities of DOPA and amyloid-forming peptides in wet conditions. Explicit solvent atomistic simulations can give insight into the interactions underlying surface adhesion in the presence of water, which can then be used to guide the design of DOPA-amyloid conjugates for underwater adhesion.

Motivated by the concern that DOPA groups might cluster in a way that interferes with their ability to adsorb to surfaces, we investigated the behavior of DOPA monomers in water. In this study, it was found that DOPA was unlikely to cluster significantly at concentrations lower than 1.0 M. DOPA-containing chains (alternating DOPA and glycine groups) were then simulated in water to examine the intra-molecular interactions of the chain, wherein we found that there were unlikely to be interactions detrimental to the adhesion process. We also simulated DOPA molecules on a silica surface and found that they readily adsorb onto silica.

With this understanding of DOPA's behavior under water and at silica surfaces, we began designing a DOPA-amyloid conjugate capable of generic underwater adsorption to varied surfaces. The amyloid-forming peptide KLVFFAE was chosen for the initial design as it is known to self-assemble into functional amyloid nanostructures. Simulations of the conjugated DOPA-amyloid chains showed that the hydroxylated phenyl rings on the DOPA motifs did not interact with the rest of the chain, which leaves them free to bind to the surface. Future work will focus on furthering understanding of the behavior of DOPA-amyloid chains at model surfaces and how changes in the backbone of the DOPA-containing chain and in the sequence of the amyloid-forming peptide affect adhesion on the model surfaces.

Copyright © American Institute of Chemical Engineers. All rights reserved.

563090 Optimizing Antifreeze Protein Activity Via an Elitist Genetic Algorithm: A Comparison of High and Low Activity Variants

Given the existence of a suitably efficient fitness function, genetic algorithms offer a promising method for systematically optimizing the properties of biological materials. In our recent paper [1], we established a neural network based method for predicting the antifreeze activity of different proteins from molecular dynamics simulation. Using this method as our fitness function, we designed a simple, but effective, elitist genetic algorithm for selecting amino acid mutations that should increase the activity of native antifreeze proteins. We apply this algorithm to two antifreeze proteins: (1) a low activity Type III antifreeze protein from the winter flounder, and (2) a hyperactive antifreeze protein from the longhorn beetle. We show that the algorithm finds structures with a predicted freezing point depression 20 - 200% greater than that of the native protein. We hope that these results inspire further investigation of high activity antifreeze biologics that could someday be used in more effective tissue preservation.

[1] Kozuch, D. J., Stillinger, F. H., & Debenedetti, P. G. (2018). Combined molecular dynamics and neural network method for predicting protein antifreeze activity. Proceedings of the National Academy of Sciences, 115(52), 13252–13257

Copyright © American Institute of Chemical Engineers. All rights reserved.

564405 Development of Polymeric Vehicles for Gene Delivery Using High-Throughput Experimentation and Statistical Learning

Complexity is a hallmark of interactions between synthetic materials and the biological environment, but frequently poses barriers to the hypothesis-guided discovery of new materials. Developing polymers for gene delivery is one such multi-dimensional design challenge that requires an intricate balance between intermolecular forces implicated in the packaging, cellular uptake and release of genetic payloads from polymer-nucleic acid complexes. Lacking a complete mechanistic picture of the effects of polymer composition and architecture on its efficiency and toxicity, polymer chemists rely on iterative "one-polymer-at-a-time" approaches that are often plagued by trial and error. Though gene editing has made rapid progress, the paucity of chemically defined, safe and efficient delivery vehicles has hindered its clinical translation. To accelerate polymer discovery for vital clinical applications such as gene editing, we need to replace traditional iterative processes with alternative experimental strategies that leverage advances in parallel synthesis and data analytics. Therefore, we hypothesize that combinatorial polymer libraries, high-throughput experimentation and statistical models can be combined to yield a powerful predictive platform for the discovery and optimization of polymeric vehicles for CRISPR/Cas9 delivery. In this work, a chemically diverse library of well-defined statistical copolymers was synthesized through RAFT polymerization and physicochemical properties such as polymer composition, molecular weight, ς-potential, pK_a, polyplex diameter evaluated. Polymer candidates were screened after complexation with two types of clinically relevant genetic payloads: 1) plasmid DNA and 2) ribonucleoprotein complexes (RNP) comprising a single guide RNA (sgRNA) and Cas9 protein. We suggest that the hit polymers identified through this study could not have been accessible through human intuition and hypothesis-driven approaches. In addition, we probed differences between polymer design criteria for plasmids and RNPs. Using the data obtained through this high-throughput screen, we expect to apply statistical techniques to mine structure-function correlations and predictive relationships that can aid experimentalists engaged in polymer design and synthesis for diverse gene therapy applications.

Copyright © American Institute of Chemical Engineers. All rights reserved.

574920 The Effect of Heteromultivalency on Lectin-Glycan Binding Process and the Identification of Critical Parameter

Glycans are the most diverse fundamental cellular macromolecules that form complex patches over the cell surfaces and bind with lectins (i.e., glycan-binding protein) [1]. The glycan-lectin binding process is important as a wide range of microbial reactions such as bacterial adhesion, viral infection, and cell differentiation, are involved with this [1-3]. One of the unique characteristics of the glycan-lectin binding process is that a single lectin often contains multiple binding pockets interacting with multiple glycan molecules (i.e. multivalency) [1,4]. We recently found that the two dimensional diffusion of glycan molecules on a cell membrane allows a homo-oligomeric lectin simultaneously binds to different types of glycans. This hetero-multivalent binding process alters the lectin binding behaviors, including binding kinetics, energy, and capacity. A living organism may utilize the hetero-multivalency to control the downstream biochemical reactions. In order to understand the role of heteromultivalency in biological processes, a new theoretical approach is needed to model this complex process.

We developed a kinetic Monte Carlo (kMC) simulation, which only relies on the fundamental physics/chemistry principles, to model the process of lectin binding to glycans on cell surfaces. The kMC simulation only considered three universal principles: binding, unbinding and diffusion [5-6]. The kMC simulation allows us to monitor the heteromultivalent binding process and the variation of binding configurations in detail. We observed that the high-affinity glycan ligands can facilitate lectin binding to other low-affinity glycan ligands on a cell membrane. Even though these low-affinity ligands are barely detectable in microarrays where ligands are immobilized, the two dimensional ligand diffusion could activate the interactions between a bound lectin and low-affinity ligands, leading to the changes in binding behaviors. To explore the critical parameters influencing heteromutivalency, we varied the glycan affinity, density, and diffusion rate in the simulations. We observed that a threshold density of low-affinity glycan ligands is required to trigger heteromultivalent binding process. Our work explains why lectin functions sometime are controlled by low-affinity ligands. In contrast to molecular dynamic simulation, the kMC simulation that can model a longer time scales (up to hours) is a better method to study the influence of relatively slow dynamics of glycan diffusion.

References

[1] Varki A. Biological roles of glycans. *Glycobiology*, **2017**, *27*, 3-49.

[2] Branson TR.; McAllister TE.; Garcia-Hartjes J.; Fascione MA.; Ross JF.; Warriner SL.; Wennekes T.; Zuilhof H.; Turnbull WB. A proteinbased pentavalent inhibitor of the cholera toxin B-subunit. *Angew. Chem. Int. Ed. Engl.*, **2014**, *53*, 8323–8327.

[3] Worstell NC.; Singla A.; Saenkham P.; Galbadage T.; Sule P.; Lee D.; Mohr A.; Kwon JSI.; Cirillo JD.; Wu HJ. Hetero-multivalency of Pseudomonas aeruginosa lectin LecA binding to model membranes. *Sci. Rep.*, **2018**, *8*, 8419.

[4] Worstell NC.; Singla A.; Wu HJ. Evaluation of hetero-multivalent lectin binding using a turbidity-based emulsion agglutination assay. *Colloids Surf. B Biointerfaces.*, **2019**, *175*, 84–90.

[5] Fallahi-Sichani M.; Linderman JJ. Lipid raft-mediated regulation of G-protein coupled receptor signaling by ligands which influence receptor dimerization: A computational study. *PLoS One.*, **2009**, *4*, e6604.

[6] Lee D.; Mohr A.; Kwon JSI. Kinetic Monte Carlo modeling of multivariant binding of CTB proteins with GM1 receptor. *Comp. Chem. Eng.*, **2018**, *118*, 283-295.

Copyright © American Institute of Chemical Engineers. All rights reserved.

580544 Using First-Principles Multi-Scale Modeling to Design Rheology Modifiers for Waterborne Paints

Modern paints are complex mixtures comprised of multiple components such as water, pigments, latex binders, dispersants, rheology modifiers or thickeners, and others. Today, Dow is a premiere supplier of binders, dispersants, thickeners, and other additives, using high-throughput experimentation and computational modeling for new formulation and process design. This presentation is specifically devoted to the use of modeling to understand the rheology and colloidal stability of a specific system -- hydrophobically ethoxylated urethanes (HEUR) thickeners in waterborne paints. Since their invention in 1970s, HEURs have been actively used as rheology modifiers for paints. Thermodynamic and rheological behavior of HEUR molecules in aqueous solutions is now very well understood and is based on the concept of transient network (TN), where the association of hydrophobic groups into networks of flower micelles causes viscosity to increase dramatically as function of polymer concentration. The behavior of complex mixtures containing water, HEUR, and latex ("binder") particles, however, is understood less well, even though it has utmost importance in the paint formulation design. To model the colloidal stability and rheology of these systems, Dow partnered with the University of Michigan research team led by Prof. Ronald Larson. We developed a multi-scale modeling framework in which the molecular details of the HEUR polymers are used to describe their adsorption onto the binder particles and ultimately the viscoelastic behavior of the overall binder/HEUR aqueous dispersion. The model is able to qualitatively describe many important features of the water/latex/HEUR mixtures. The proposed approach could potentially lead to the design of new HEUR structures with improved rheological performance. Collaboration: Antony van Dyk, Alan Nakatani, Tirtha Chatterjee, Susan Fitzwater, Alex Kalos (Dow); Ronald Larson, Fang Yuan, Shihu Wang, Hussein Rezvantalab, Elnaz Hajizadeh (University of Michigan).

Copyright © American Institute of Chemical Engineers. All rights reserved.

580545 Applications of Computational Chemistry in the Development of Materials for the Semiconductor Industry

As the size of semiconductor devices keeps going down and the complexity of the architecture keeps going up, the demand for advanced specialty materials to deposit, etch, polish, clean, package, etc. the wafers during manufacturing is also increasing as the nodes advance. The materials used represent a wide variety of chemistries ranging from small molecules to complex liquid phase formulations while covering a significant portion of the periodic table. In order to shorten the development cycle of new materials as well as improve the general understanding of the underlying mechanisms of the different processes, computational chemistry (often supported by additional modeling tools) have been found useful and is widely practiced. Depending on the system to be studied, the computational approaches need to be flexible. Model systems range from small gas phase molecules treated by highly accurate traditional ab initio methods and polymeric systems characterized by molecular mechanics, to complex multiphase systems for which QSAR approaches could be the most useful. Illustrative examples are presented to show real life issues along with the contribution of computational chemistry to the solution. The examples are specific to advanced materials used in the semiconductor industry.

Copyright © American Institute of Chemical Engineers. All rights reserved.

580542 Molecular and Detailed Isotopic Structures of Petroleum: Kinetic Monte Carlo Analysis of Alkane Cracking

The rapidly growing field of "clumped isotopes" chemistry provides analytical detail that goes beyond identifying the abundance of a single rare isotope within a molecule. Rare isotopes may prefer certain non-equivalent positions in a molecule and multiple rare isotopes within a single molecule may have a tendency to clump together. Accurate and precise measurement of these features has enabled applications such as new geochemical thermometers, providing temperatures of past environments, and new abilities to trace the origin of species found in the atmosphere. Developing additional applications depends on our being able to link the structure of target species with the structure of their source molecules and with all the current and past processes influencing the detailed isotopic structure of the molecules.

In the specific example of interest here, developments in clumped isotope geochemistry have given rise to a suite of measurements that provide new signatures related to the thermal history of low molecular weight hydrocarbon gases found in sedimentary basins. However, to date no study has linked these clumped isotope features, along with traditional compositional and bulk isotopic measurements, to any mechanistic understanding of how these signatures develop and evolve. In this study, we developed a kinetic Monte Carlo method to predict consistent and simultaneous molecular distributions, bulk isotopic content, and detailed (multiply substituted and site-specific) isotopic structures of hydrocarbons from a hydrocarbon cracking model. The detailed isotopic structure of the source hydrocarbons (initially modeled as long alkanes), the intermediates, and the product molecules is followed as a function of the level of conversion due to cracking reactions. The bulk 13C content of gaseous alkane products generated via the model is shown to follow the linear natural gas "Chung" plot at low conversion, but deviates at higher degrees of conversion. As examples of the information generated, the populations of center vs. terminal 13C-substituted propane are reported as a function of the starting alkane chain length and level of conversion and the population of doubly-13C-substituted ethane is described as a function of the level of conversion and for different 13C substitution patterns in the source hydrocarbons. The results are compared to experimental data where possible and highlight the possibility of constraining hydrocarbon source isotopic structure and the nature of the generation processes leading to the formation of natural hydrocarbon deposits.

Copyright © American Institute of Chemical Engineers. All rights reserved.

580541 Improving utilization of quantum chemistry in industrial research

Quantum chemistry as a tool has achieved wide-spread acceptance and utilization in academic research environments; however, these are significantly less in the industry with the exception of a few leading companies. We consider that the most important reasons are:

1. a) the lack and speed of developing detailed mechanistic understanding of the underlying chemistry of processes
2. b) availability and access to experts-developed computational work-flows for accurate prediction of specific molecular properties
3. c) rapid search, access, and presentation of computed information and
4. d) translation of those to engineering modelling tools.

We present application examples where improvements of these greatly helped to achieve success. The methods include mechanism, species and their conformers generation, large-scale computations, data-management, data-analysis for validations and user interfaces. The underlying tools, besides a few commercial computational chemistry packages, utilize publicly available cheminformatics, database, and data analysis/modelling software. The developed interfaces and computational workflows enable not only experts of quantum chemistry but also other fields' to efficiently access and utilize its best of class methods in research, and fast-paced production problem solutions.

Copyright © American Institute of Chemical Engineers. All rights reserved.

580543 Industrial Applications of Materials Informatics

In recent years, the application of Materials Informatics in Materials Science-based industrial research has been increasing at a rapid pace. Advances in computational hardware, software algorithms, and data management have made this possible. In addition, industrial companies are combining the skill sets of materials scientists and engineers, chemists, chemical engineers, data scientists and additional disciplines as projects warrant to enable more rapid progress. Several examples of the application of Materials Informatics in Materials Science-based new product development will be presented.

Copyright © American Institute of Chemical Engineers. All rights reserved.

574499 Computing Chemical Reaction Equilibria in Molten Salts from Molecular Simulations: Composition of Molten Carbonate Electrolytes in Contact with Gaseous Species

Several common molecules including water, oxygen, and carbon dioxide have significant solubility in molten salt phases. Upon absorption, these molecules chemically react with the molten salt to form a diversity of species. For example, dissolved oxygen in molten alkali-metal carbonates is believed to exist almost entirely as a combination of peroxide and superoxide species. The precise ratio of these "activated oxygen species" (OASs) depends on several factors including the metal-cation composition and melt acidity. The solvation of CO_2 and H_2O leads to similar complex reaction products. The interplay between dissolved gas composition and factors such as temperature and melt cation composition is only poorly understood. Molecular simulations can provide insight into structure-function relationships, which can then guide rational design of improved electrolytes. However, determining chemical reaction equilibria (CRE) in condensed phases, especially for complex molecular fluids like molten salts, is computationally challenging. In this talk we describe our work developing methods for calculating CRE for salts and solutes relevant for Molten Carbonate Fuel Cells. Our methods utilize stepwise ion insertions to calculate the chemical potentials of all species combined with quantum chemical methods to determine gas-phase thermochemical data when needed. We show our first calculations of equilibrium concentrations of CO_2, O_2, and H_2O, and their activated products, in Li/K and Li/Na molten carbonate electrolytes. Finally, we discuss how changes in CRE can be related back to physical properties such as entropy and polarizability of alkali-metal cations, and transport of electrochemical species.

Copyright © American Institute of Chemical Engineers. All rights reserved.

580342 Controlling the Properties of Molecular Electrolytes

The physical properties of molecular electrolytes are essential for biological function and of great interest to industry. They have been particularly difficult to describe theoretically and computationally given the presence of composition and dielectric heterogeneities and the large number of disparate length scales involved in determining their physical properties. The Debye length, the Bjerrum length, the ion size, the molecular configuration, and the charge distribution along the molecules determine their structure and their response to external stimuli. Computational, numerical and theoretical approaches to describe electrolytes will be described and a conceptual understanding of how charge dictates nano-scale structure and function will be provided.

Copyright © American Institute of Chemical Engineers. All rights reserved.

572040 Developing SAFT-gamma Mie Models for Aqueous Electrolytes: Organic Salts and CO_2-Brine Mixtures

Thermodynamic modelling of electrolyte solutions proves particularly challenging for complex systems characterized by the presence of organic electrolytes, which are generally non-spherical molecules. SAFT-γ Mie [1,2] is a successful statistical thermodynamic framework to model bulk thermodynamic properties. SAFT-γ Mie is essentially the SAFT-VR Mie equation of state casted in a group contribution formulation, where each component can be constructed using functional groups that represent various molecular moieties. SAFT-VR Mie has been recently used to study ionic solutions [3] by incorporating the Mean Spherical Approximation (MSA) approach to account for the Coulombic interactions between ions in the presence of solvent molecules; the polar-polar and ion-polar interactions are implicitly treated with a separate Born solvation energy term. However, both of these electrostatic terms are limited to spherical molecules and further approximations are required to treat non-spherical species.

Molecular simulations have been widely used to assess the necessary approximations to develop statistical mechanical theories of fluids. In this work we compare the electrostatic contributions to the chemical potential predicted using various primitive approaches (MSA, Debye-Huckel, Born) against simulation data for model systems that incorporate the solvent explicitly. We also compare two different methodologies that we developed to account for cases were some of the charged compounds are non-spherical.

Estimating the chemical potential of charged molecules using traditional molecular simulation techniques is challenging. Thus, we have used the expanded ensemble transition matrix method [4] to run these calculations robustly and efficiently. The theoretical predictions were compared with simulation data to provide a better understanding of the implications of the assumptions employed within the SAFT-γ Mie electrolyte framework.

Finally, the methodology is applied to real systems like brine-CO_2 mixtures and short chain carboxylate salt solutions. The fitted models for these mixtures are characterized by good accuracy and can be used to model properties like salt and CO_2 solubility, vapor pressures and densities.

References

[1] V. Papaioannou, T. Lafitte, C. Avendaño, C.S. Adjiman, G. Jackson, E.A. Müller and A. Galindo, *J. Chem. Phys.*, 140, 054107 (2014)

[2] V. Papaioannou, F. Calado, T. La, S. Dufal, M. Sadeqzadeh, G. Jackson, C.S. Adjiman, A. Galindo, *Fluid Phase Equilib.*, 416, 104 (2016)

[3] D.K. Eriksen, G. Lazarou, A. Galindo, G. Jackson, C.S. Adjiman, A.J. Haslam, *Mol. Phys.* 114, 2724 (2016)

[4] A.S. Paluch, S. Jayaraman, J.K. Shah, E.J. Maginn, *J. Chem. Phys.* 133 124504 (2010)

Copyright © American Institute of Chemical Engineers. All rights reserved.

574199 A Universal Equation of State for Hard Polyhedra

The empirical formulation of the ideal gas law marked a major turning point in the development of theoretical models aimed at describing fluid behaviors. Its extension to real systems fails due to the presence of both excluded volume and attractive interactions in non-ideal particles. Consideration of particle shape further complicates the development of a universal theoretical framework, although there have been several attempts with varying degrees of accuracy for convex shapes. Here, we introduce a new equation of state for hard polyhedra. The derived expression not only shows excellent agreement with simulations, but also suggests a "corresponding state"-like behavior, allowing for a reduction across shape space onto a single master curve as well as prediction of the order-disorder transition density for polyhedra

Copyright © American Institute of Chemical Engineers. All rights reserved.

572856 Systematic Evaluation of Thermophysical Properties of Low GWP Refrigerants and Blends for Refrigeration Cooling Systems By Molecular Modeling Techniques

According to the Paris Agreement [1], a scenario in which the global average temperature is not more than 2°C from pre-industrial levels could significantly reduce the risks and impacts of climate change. As mentioned by the International Energy Agency, the global energy-related CO_2 emissions in 2017 reached a historic high of 32.5 $GtCO_2$ [2], value that have to be reduced below 20 Gtonne-CO_2 per year by 2050, and near-zero emissions should be reached by 2100 [3]. Different technologies need to be developed or improved in order to mitigate climate change, among them, the development of environmentally friendly refrigerants. Commonly used refrigerants such as chlorofluorocarbons (CFC) and hydrochlorofluorocarbons (HCFC) were found to be ozone depleting by the mid-1980s. A global decision to phase out ozone depleting such substances, the Montreal Protocol, was successfully made in 1987. Consequently, the application of hydrofluorocarbons (HFC) as alternative for refrigeration units grew over the past couple of decades. However, HFCs are known to be greenhouse gases (GHGs) with considerable GWP, thousands of times higher than carbon dioxide [4]. Kigali's Amendment to the Montreal Protocol [5], a global deal entering into force on 2019, legally binds world's nations to gradually phase out the consumption and production of HFCs by the late 2040s. This Amendment has promoted an active area of research towards the development of low GWP refrigerants, and will significantly contribute to the Paris Agreement by avoiding nearly half a degree Celsius of temperature increase by the end of this century [5].

Most of the currently in place and proposed regulations (F-Gas in the European Union and CARB and EC in North America) [6] target refrigerants to be used that do not exceed 150 GWP for refrigeration and 750 GWP for air conditioning applications, thus creating an immediate demand for developing fourth generation refrigerants with low GWPs. The major perceived barriers to these low-GWP alternatives include safety (i.e., flammability, corrosion, toxicity), first cost and return on investment. In response to this need, hydrofluoroolefins (HFOs) and some third generation refrigerants blends have been included in the NIST list as the most environmental friendlier candidates found so far, with GWP values comparable to those of hydrocarbon (HC) based refrigerants.

Refrigerants are complex molecules and their blends are highly non-ideal mixtures. Understanding the phase behaviour and accurately predicting the thermophysical, interfacial and transport properties of these systems is essential for designing and evaluating refrigeration cycle performances and determining the optimal compositions [7,8]. Hence, the development of new refrigerants requires not only a low GWP, but also an accurate evaluation of their physicochemical properties to determine their capacity to substitute HFCs refrigerants without a loss of efficiency. As mentioned, HFOs represent a realistic alternative due to their excellent environmental properties, although its relative high flammability prevents their use as single compounds. Hence, current options consider the combination of HFCs or HCs with HFOs, forming azeotropic blends with similar properties but lower GWPs [9,10].

We will present and discuss here results obtained from a consistent thermodynamic model that characterizes new fourth-generation HFOs-based refrigerants with the SAFT approach [11], using the soft-SAFT [12] model compared with experimental data and with (Polar) PC-SAFT [9,10,13].

The soft-SAFT evaluation of the HFOs has been performed based on a molecular model transferred from HFCs, which had already been characterized in a previous work [4], taking advantage of the similarities between the two chemical structures. As a result, a set of optimized molecular parameters for these HFOs is provided, being consistent from a structural perspective. The description of the saturated density and vapor pressure is very accurate, in all cases, while the heat capacity and the speed of sound are predicted to validate the model, finding also very good agreement with the experimental information. Additionally, a very good description of the surface tension and the viscosity of these compounds is achieved by using the Density Gradient Theory and the Free-Volume theory coupled into soft-SAFT. Next, blends between HFCs and the two most common HFOs, R1234yd and R1234ze have been adequately described, providing a realistic picture of their phase equilibria, interfacial behavior and viscosity. The model can then be used for the prediction of any blend by varying the composition, as well as the behavior of ternary mixtures of refrigerants.

The work is completed with a process simulation of a cascade refrigeration system, where the common 3rd generation refrigerant R410 is compared with these new 4th generation mixtures. The thermodynamic information of the system is calculated using the soft-SAFT model and introduced in the process simulator Aspen Plus. A comparison of the Coefficient of Performance (CoP) is done for different cases, in order to establish the best alternative to R410.

Acknowledgements

This research is supported by Project KET4F-Gas – SOE2/P1/P0823 which is co-financed by the European Regional Development Fund within the framework of Interreg Sudoe programme. Additional partial support has been provided by Khalifa University of Science and Technology under project CIRA-2018-121.

References

1. https://unfccc.int/sites/default/files/english_paris_agreement.pdf
2. Global Energy & CO_2 Status Report 2017, https://www.iea.org/publications/freepublications/publication/GECO2017.pdf, accessed December 2018
3. Climate Change 2014: Mitigation of Climate Change. Cambridge University Press; 2014.
4. Metz B, Davidson O, de Conick H, Loos M, Meyer L. 2005. IPCC. Cambridge University Press, UK. (https://www.ipcc.ch/report/carbon-dioxide-capture-and-storage/ , accessed December 2018).
5. United Nations treaty Collection. Amendment to the Montreal Protocol on Substances that Deplete the Ozone Layer. Kigali, 15 October 2016 (https://treaties.un.org/Pages/ViewDetails.aspx?src=IND&mtdsg_no=XXVII-2-f&chapter=27&clang=_en, accessed December 2018).
6. Graziosi F, Arduini J, Furlani F, Giostra U, Cristofanelli P, Fang X, et al. European emissions of the powerful greenhouse gases hydrofluorocarbons inferred from atmospheric measurements and their comparison with annual national reports to UNFCCC. Atmospheric Environment. 2017;158:85-97.
7. Qi Y, Zhang H, Zhao W, Liu Y, Liu X. International Refrigeration and Air Conditioning Conference. Purdue University. 2014.
8. Kim M-H, Pettersen J, Bullard CW. Fundamental process and system design issues in CO_2 vapor compression systems. Progress in Energy and Combustion Science. 2004;30(2):119-74.
9. Fouad WA, Vega LF. The phase and interfacial properties of azeotropic refrigerants: the prediction of aneotropes from molecular theory. Physical Chemistry Chemical Physics. 2017;19(13):8977-88.
10. Fouad WA, Vega LF. Transport properties of HFC and HFO based refrigerants using an excess entropy scaling approach. The Journal of Supercritical Fluids. 2018;131:106-16.
11. Chapman WG, Jackson G, Gubbins KE. Phase equilibria of associating fluids. Molecular Physics. 1988;65(5):1057-79.
12. Blas FJ, Vega LF. Thermodynamic behaviour of homonuclear and heteronuclear Lennard-Jones chains with association sites from simulation and theory. Molecular Physics. 1997;92:135-50.
13. Gross J, Sadowski G. Perturbed-Chain SAFT: An Equation of State Based on a Perturbation Theory for Chain Molecules. Industrial & Engineering Chemistry Research. 2001;40(4):1244-60.
14. Vilaseca O, Llovell F, Yustos J, Marcos RM, Vega LF. Phase equilibria, surface tensions and heat capacities of hydrofluorocarbons and their mixtures including the critical region. J. Supercritical Fluids 2010:55 (2): 755-768.

Copyright © American Institute of Chemical Engineers. All rights reserved.

577068 Systematic Development of an Artifact-Free SAFT Eos

Recent work by Jaubert, Polishuk, Alsaifi, and coworkers has demonstrated several artifacts in the most popular implementations of SAFT EOSs. For example, Jaubert and coworkers showed that the PC-SAFT EOS trends to negative critical pressures in the long chain limit and indicates an artificial pure component liquid-liquid separation and critical point at high densities.[1] Polishuk addressed the high density artifacts by regressing new coefficients for PC-SAFT, but his proposed EOS gave poor results for vapor pressures of heavy n-alkanes.[2] Alsaifi et al. developed a "bifurcation" algorithm that mapped artifacts throughout the T-ρ plane, demonstrating artifacts in HR-SAFT, Soft-SAFT, and SAFT-VR at intermediate densities and temperatures.[3] On the other hand, the cubic SAFT formulation of the ESD EOS[4] and the multiparameter formulation of the SPEAD11 EOS[5] were free of artifacts at $T > 40$K. Artifacts can be quite problematic when solving for global free energy minima during phase equilibrium calculations of complex mixtures. Artificial density roots inside the binodal envelop also cause classical density functional theories for interfaces to fail. Altogether, these observations suggest a need and an opportunity for a protocol by which artifacts can be avoided and robust SAFT EOSs provided.

To develop a systematic protocol, we begin by revisiting the ESD EOS to examine how a cubic SAFT EOS can be developed while matching molecular simulation results developed since 1990. We find that the denominator of the repulsive contribution can be improved by recognizing that heavy branched hydrocarbons may remain as liquid at much higher packing fractions than spherical molecules. This suggests generalizing the factor of 1.9 to a function of molecular weight and functionality. Analysis of the attractive contribution shows that agreement with simulations of long chains can be improved by eliminating the temperature dependence of the denominator, resulting in an EOS that is practically first order in temperature. The resulting cubic SAFT EOS is shown to be free of artifacts at all temperatures, densities, and chain lengths.

The next step is to add features to the cubic EOS that enhance accuracy relative to molecular simulations while retaining freedom from artifacts. Naturally, the resulting EOS is no longer cubic. Analyzing the derivative behavior similar to the SPEAD11 EOS analysis leads to identification of several constraints that must be satisfied when characterizing the higher order contributions to thermodynamic perturbation theory (TPT). These constraints are much easier to satisfy when the higher order TPT contributions decay monotonically from their critical value to zero at high density, as in the Weeks-Chandler-Andersen (WCA) split of the potential function. An infinite order EOS is demonstrated that is free of artifacts while reproducing simulation results for Lennard-Jones chains with reasonable accuracy.

Finally, a simplified association contribution is added to the chain EOS to examine whether artifacts may arise in the presence of association. This leads to a modification of the constraints developed for non-associating molecules. The protocol is complete in cases where the WCA split of the potential function is acceptable.

References

[1] R. Privat, R. Gani, and J. Jaubert, "Fluid Phase Equilibria Are safe results obtained when the PC-SAFT equation of state is applied to ordinary pure chemicals ?," vol. 295, pp. 76–92, 2010.

[2] I. Polishuk, "Standardized critical point-based numerical solution of statistical association fluid theory parameters: The perturbed chain-statistical association fluid theory equation of state revisited," *Ind. Eng. Chem. Res.*, vol. 53, no. 36, pp. 14127–14141, 2014.

[3] N. M. Alsaifi, M. Alkhater, H. Binous, I. Al Aslani, Y. Alsunni, and Z. G. Wang, "Nonphysical Behavior in Several Statistical Mechanically Based Equations of State," *Ind. Eng. Chem. Res.*, vol. 58, no. 3, pp. 1382–1395, 2019.

[4] J. R. Elliott, S. J. Suresh, M. D. Donohue, and M. D. Donohue, "A Simple Equation of State for Nonspherical and Associating Molecules," *Ind. Eng. Chem. Res.*, vol. 29, no. 7, 1990.

[5] A. F. Ghobadi and J. R. Elliott, "Evaluating perturbation contributions in SAFT models by comparing to molecular simulation of n-alkanes," *Fluid Phase Equilib.*, vol. 306, pp. 57–66, 2011.

Copyright © American Institute of Chemical Engineers. All rights reserved.

579175 Liquid-Liquid Equilibria for Ternary Systems Limonene + Linalool + Ionic Liquid with Lactate Anion

Essential oils are important in the food and cosmetic industries because of their characteristic tastes and odors. They are composed of one or more terpenes and some oxygenated derivatives. The oxygenated compounds can be oxidized when they are exposed to air, and they usually have organoleptic properties. Separation is required to concentrate these oxygenated compounds. Liquid-liquid extraction would be more suitable than distillation, because application of heat can affect the organoleptic properties of essential oils. In this work, we focus on liquid-liquid extraction as a separation method for essential oils.

Solvent selection is very important in the design and development of the liquid-liquid extraction processes. Recently, alternative solvents for green chemistry have been attracted much attention as promising candidates to replace toxic organic solvents and volatile organic compounds (VOCs). We focus on ionic liquids (ILs) with bio-based anion. ILs with bio-based anion have little effect of environment and life cycle. Several building blocks derived from major renewable sources can be used as IL precursors. Because of their features, ILs with bio-based anion could be used as a solvent for green separation and purification. In the design and development of liquid-liquid extraction processes. Reliable liquid-liquid equilibrium (LLE) data are of importance

This work presents the experimental LLE of ternary mixtures limonene + linalool + IL at temperatures 298.15 and 323.15 K. Two ILs with bio-based anion, i.e., 1-butyl-3-methylimidazolium lactate [BMIM][lac] and 1-hexyl-3-methylimidazolium lactate [HMIM][lac] were tested. From the experimental LLE data, considerations of two ILs as extraction solvent were done by the values of distribution factor (β) and selectivity (S). From their considerations, both ILs could be more suitable and green solvents for the separation of deterpenation of citrus essential oil using liquid-liquid extraction.

Copyright © American Institute of Chemical Engineers. All rights reserved.

578324 Equation of State of the One-Dimensional Lennard-Jones System: Thermodynamics and Fluctuations

The thermodynamic properties of one-dimensional systems are of interest for several reasons. Exact analytical solutions are, in some cases, obtainable for 1D systems even when the 3D analogue has no such solution. Those solutions can aid in identifying physics and optimum numerical approximation methods in the 3D analogue systems. However, 1D systems also exhibit surprising behaviors not observed in higher dimensions, which are also interesting subjects of research, such as the lack of a phase transition (in the absence of infinitely long-ranged interactions). And, perhaps most interestingly, quasi-1D nanomaterials (nanotubes, nanoribbons, nanowires, etc.) are rapidly gaining interest for a variety of technological applications, and 1D models may serve as a useful first-order approximation or reference state for modeling these materials, in a manner similar to the roles played by, for example, the ideal mixture or the hard sphere fluid when studying real fluids. However, the study of 1D systems is complicated by large fluctuations in density and energy, long-range correlations, and long relaxation times.

The Lennard-Jones pair potential is a simple and familiar interaction model, and thus, a good place to begin studying the properties of 1D systems. We have applied John A. Barker's general equation of state for 1D systems of classical particles to obtain the equation of state of the 1D Lennard-Jones fluid. In addition to presenting the density, internal energy, and entropy of this system, we consider properties dependent on first-order fluctuations, including heat capacity, compressibility, coefficient of thermal expansion, and thermal pressure coefficient. We obtain numerical solutions for these properties within a large subset of the thermodynamic phase space. We then test the resulting density and internal energy expressions using isothermal-isobaric Monte Carlo simulation, obtaining excellent agreement. We will describe the expressions obtained for these properties, our methods of numerical solution and simulation, and the results obtained.

Copyright © American Institute of Chemical Engineers. All rights reserved.

559625 How Molecular Simulations Can Inform Business Decisions in Different Industrial Sectors: The Composelector Example

The integration of modeling and simulations techniques to support material selection processes (MSPs) is one of the most compelling needs in advancing material industry and manufactory, due to the necessity of effective/efficient design and production of sophisticated materials, components and systems with advanced/extreme performance on a competitive time scale. In this arena and, specifically, for complex structural materials such as polymer-based nanocomposites (PNCs) there is a strong industrial demand for chemistry/physics based models and modeling workflows able to predict relevant material properties (aka Key Performance Indicators or KPIs) in an accurate and reliable way. With the aim of filling the gap between business processes and materials science/engineering workflows, this work presents the application – within the framework of the EU H2020 project COMPOSELECTOR – of a multi-disciplinary, multi-model approach for the accurate, reliable, efficient and cost-effective industry-driven KPIs determination for PNC materials.
Specifically, the results obtained for case studies pertaining to aerospace (AIRBUS S.A.S.), automotive (DOW Chemicals Europe SA) and tire (Goodyear Tire & rubber Company) applications and the relevant KPIs determination (i.e., mechanical, viscoelastic and interfacial properties) will be presented and discussed. The high degree of modeling and data prediction integration into MSPs will ultimately yield valuable information about current “holes in property space”, enable what-if scenarios and provide mechanistic insights into the high performance of these fascinating materials.

Copyright © American Institute of Chemical Engineers. All rights reserved.

560202 A Machine Learning Guided Approach for Studying Local Solvation Environments

In this work we introduce a calculation scheme that uses state of the art computational tools to study ion solvation environments. We first generate microsolvated clusters with global optimization algorithm called ABCluster and then introduce a measure of convergence based on the structural similarities of solvent molecules near the solute using unsupervised machine learning algorithms called Smooth Overlap of Atomic Positions (SOAP) and sketchmaps. SOAP is a clustering technique that identifies structures that are closely related to each other, and the sketchmaps reduce the dimensionality of the feature vectors to represent the SOAP kernels. After studying different solvation environments, we calculate solvation energies using Quasi Chemical Theory (QCT) and thermodynamic cycles. We demonstrate that low energy molecular clusters produced by our procedure have structurally similar local solvation environments, and this suggests that this calculation scheme can be used to quantify the number of explicit solvent molecules needed to accurately model the relevant local solvation environment of a charged solute. We expect this approach will be broadly applicable for the atomistic modeling of ions in different solvents.

Copyright © American Institute of Chemical Engineers. All rights reserved.

560363 Relating Segregation Behavior in Metal Alloy Nano-Particles of Different Shapes and Sizes

Surface segregation in metal alloy nano particles (NP), i.e., preferential enrichment of one of constituent metal species at the top few surface layers, is important to adsorption, catalysis, separation and optoelectronic applications. Among the large number of factors that influence segregation are NP shape and size, which can be experimentally hard to control. Therefore, an efficient computational strategy is required to estimate the segregation profiles in nano particles of a given size and shape distribution. The talk focuses on a nano-thermodynamic model based on distribution coefficient Δ, which we have developed recently. While Δ helps to determine the preference of metal species to be at surface facets, edges, vertices and bulk regions, the nano-thermodynamic model includes the number of sites in each region so that the segregation profiles in nano particles of different shapes and size can be predicted. Δ, which is a function of the alloy composition in each region, is determined for a NP of a given size and shape, and directly employed with NPs of other sizes and shapes. Different NP shapes like cube, truncated-octahedron and icosahedron are explored. Using Au-Pt, Ni-Pt and Ag-Au as prototype systems, we demonstrate that a wide range of alloying behavior from complete mixing to complete segregation can be captured in bimetallic nano particles of different shape and size using the nano-thermodynamic model.

Copyright © American Institute of Chemical Engineers. All rights reserved.

562756 Influence of Composition on Nucleation Process in CuNi System

By using molecular dynamics, we show the effect of the composition on the nucleation process. To do so, we use three different systems, pure copper, pure nickel and the copper-nickel alloy. The system is sampled with the umbrella sampling that allows us to look at its free energy. Our results show a higher free energy for the alloy than for the pure systems. Meanwhile, the alloy tends to form a bigger critical nucleus than the two other systems. The structure as well as the composition of the nucleus has also been studied showing how it has been impacted by the composition.

Copyright © American Institute of Chemical Engineers. All rights reserved.

566598 Automating Transferable Coarse-Model Generation and Back-Mapping with an Open Source Toolchain

Coarse-grained models have enabled access to length scales and time scales in molecular simulations that permit structure-property relationships to be probed in complex soft materials. For example, in previous work, we found that simulation volumes with 35nm edge lengths are necessary for to avoid finite size effects in certain toughened epoxy thermoset systems, and 2.1nm periodic features in organic photovoltaic melts require (10nm)^3 volumes to resolve. When coarse-grained models are derived from more detailed structural information, these techniques can simultaneously provide accurate views of morphology while accelerating equilibration dynamics. Reconstructing atomistic representations is often needed for precise prediction of mechanical or electrical properties. As examples, the elastic modulus of simulated epoxies and the charge transport properties of organic photovoltaic morphologies are far more reliable from atomistically-detailed representations of structure.

In this work, we present coarse-graining infrastructure built using open source molecular modeling tools. We utilize SMARTS strings to define chemistries and coarse-grained mappings. We show how this functionality enables facilities rapid prototyping of coarse models as well as high-throughput screening of multiple chemistries with varying degrees of resolution. In sum, we describe techniques for quickly generating and evaluating models that enable trade-offs between speed and accuracy to be rationally informed. We demonstrate its utility for the multiscale modeling of complex polymer systems. We conclude by discussing examples of how this toolchain can be used to cross-validate simulation predictions, enhance reproducibility, and extensibility.

Copyright © American Institute of Chemical Engineers. All rights reserved.

566616 Computational Study of Liposome Formation in Coaxial Turbulent Jet Flow Using CG-MD and CFD Approaches

Purpose: Liposomes are spherical structures consisting of single or multiple lipid-bilayer membranes that circumvent the central aqueous phase. They are used as a carrier for drug delivery systems including anti-cancer and anti-fungal drugs with several FDA-approved clinical applications. We have focused on the continuous process in which two liquid flows are controllably mixed using a jet in co-flow, of which the inner flow consists of a solution of lipids in ethanol and the exterior flow is an aqueous phase. Intermolecular forces between molecules cause lipid aggregation and, along with the jet flow characteristics, results in the formation of liposomes. However, the underlying mechanism as well as details of the process are only partially understood. We have carried out a multi-scale computational study of liposome formation in coaxial turbulent jet flow to probe the underlying mechanism and quantitatively predict liposome properties.

Methods: Both computational fluid dynamics (CFD), as a macro-scale simulation, and coarse-grained molecular dynamics (CG-MD), as a micro-scale investigation, have been conducted to not only reveal the detailed mechanism of liposome formation, but also implement multi-scale case studies for the process.

Results: CFD simulations were verified by comparison of flow pattern as well as formation temperature with experiments. Our CG-MD simulations indicate that MARTINI force field (FF) does not capture a realistic behavior of lipids and cholesterol in ethanol solution since the lipids aggregate which contradicts with the results of the experimentation and all-atom MD simulation. We used a versatile object-oriented toolkit for coarse-graining applications (VOTCA) to optimize MARTINI FF with reference to all-atom MD simulations. Our multi-scale simulations are in accurate agreement with experimental quantitative results and trends.

Conclusions: Discrete and continuum computational modeling of liposome formation cover micro-scale and macro-scale respectively which overlap in a short scale domain due to computational feasibility, therefore these two approaches are complementary to one another. Our CFD simulation shows that including heat of mixing in the energy equation is inevitable in order to obtain a formation temperature comparable with the experiments. Default MARTINI potential energy parameters require optimization with reference to an all-atom MD simulation for properly modeling lipids in ethanol solution.

Acknowledgements: FDA Grant# 1U01FD005773-01.

Disclaimer: This article reflects the views of the authors and should not be construed to represent FDA's views or policies.

Copyright © American Institute of Chemical Engineers. All rights reserved.

568396 Analyzing the Bandgap of 3D Perovskite Oxides: Machine Learning Approach

In this study we have developed a machine learning approach to analyze and predict the bandgaps of 1146 ABO_3 perovskites, computed using density functional theory. Since DFT-computed bandgaps with high accuracy are normally also highly resource demanding, we developed a machine learning model to predict the perovskite bandgaps by utilizing readily available atomic and crystal descriptors. These descriptors include atomic electrochemical properties such as electron affinity and ionization energy of the A/B-site atoms, as well as Goldschmidt tolerance factor of the crystal structure. As we arrive at our results, we conduct a comparative study of two machine learning algorithms: namely kernel ridge regression and gradient boosting regression. Along the way, we employ recursive feature elimination and correlation based feature selection methods to remove the weakest inputs from our dataset, and reduce the 35 starting features down to 28 most essential features, resulting in a final model mean square error of 0.2 eV. Lastly, we performed a relative contribution analysis of the inputs with regards to their influence on the bandgap of the perovskites. In essence, we attempt to use machine learning as an analytical technique for evaluating the structure-bandgap relationship of perovskite.

Copyright © American Institute of Chemical Engineers. All rights reserved.

568615 Investigating Hybrid Organic-Inorganic Tin Perovskites for Li-Ion Battery Applications: DFT Modeling Approach

This study highlights the application of 3D hybrid organic-inorganic perovskites (HOIPs) for use in lithium-ion battery electrode applications. Demand for more efficient electrochemical storage materials is rising year by year in a wide range of practical applications. HIOPs have attracted intensive attention for many applications mainly due to the highly versatile nature of perovskites as well as their solution-processability at low temperatures. The use of perovskite in Li-ion batteries has only been recently examined as most studies so far have focused on the use of HOIPs for optoelectronic applications. Specifically, most recent studies have examined the use of lead based HOIPs, which possess a number of possible environmental concerns due to the toxicity of lead. Thus, in this study, we use DFT to investigate lithium intercalation in alternative tin based perovskites ($ASnX_3$, where A = MA^+, FA^+, and Cs^+ and X = Cl^-, Br^-, and I^-). We characterize the performance of these perovskites for Li-ion applications by considering the intercalation voltage for lithium as well as the structural distortions incurred by the SnX_6 octahedra.

Copyright © American Institute of Chemical Engineers. All rights reserved.

568746 Statistical Learning Approaches for Identifying Descriptors of Metal/Support Interactions in Catalysis

In this work, we apply statistical learning (SL) methods to identify physical descriptors for predicting changes in metal/oxide interactions that occur in heterogeneous catalysis. The binding energy of transition metal atoms on oxide surfaces can be influenced by the presence of adsorbates or surface impurities that alter the electronic composition of the oxide support. We use SL approaches to predict the extent to which these surface-modifiers alter metal/support interactions based on readily available physical properties of the supported metal and the oxide. Least absolute shrinkage and selection operator (LASSO) is a well-established tool for feature selection in catalysis and material science. However, a persistent flaw of LASSO is its high rate of false positives, where features are selected for the model that are not true independent descriptors of the data in the training set. To combat this issue, we introduce two additional Bayesian methods, Horseshoe priors and Dirichlet-Laplace priors, in our feature selection process. Unlike LASSO regression, these two Bayesian methods estimate a distribution of feature coefficients and the corresponding error. These distributions then are updated using iterative Markov chain Monte Carlo (MCMC) sampling until the distribution reaches an equilibrium prediction. The descriptors with coefficients greater than a pre-defined cutoff threshold then are selected as the best descriptors for the final model. Well-selected priors in these Bayesian methods reduce false positives, thus leading to more robust descriptor sets for general models of metal/support interactions in catalysis.

Copyright © American Institute of Chemical Engineers. All rights reserved.

568809 Multi-Scale Simulations with Linear Scaling Relations and Microkinetic Models of Aqueous Phase Reforming of Sugar Alcohols

With rising interest in the bio-refinery industry comes interest in converting biomass to useful chemicals and fuels. Specifically, there is interest in converting sugar alcohols, such as glycerol, methanol, ethanol and ethylene glycol, some of which currently exist in surplus. Currently, techniques exist to convert these sugar alcohol species to H_2, CO, and CO_2 gas, as well as liquid products such as acids and ketones.

Among those techniques, aqueous phase reforming features low operation temperatures. Further, it is carried out in aqueous conditions, which are the same conditions as the sugar alcohols in surplus would be supplied. However, challenges in catalyst cost, selectivity and activity impede the broader expansion of this process. Design of new catalysts requires substantial knowledge of the reaction mechanism and operation conditions, which is challenging due to the large reaction networks, consisting of hundreds to thousands of possible reactions, and the aqueous environment itself, which complicates the mechanistic steps and also impedes experimental and computational attempts at observing the catalysis. In this work, we develop computational methods to overcome these challenges. To understand the influence of the water environment on the energetics and mechanism, we develop a multi-scale simulation strategy that employs force-field molecular dynamics (FFMD) and density functional theory (DFT). In this strategy, DFT is used to optimize surface structures and calculate catalytic energies, and FFMD is used to simulate the liquid water solvation. This strategy enables the capture of both energetic and mechanistic influences of water on the catalysis. To overcome the large number of potential reactions, we derive scaling relations, which are linear equations that correlate thermodynamic and kinetic properties of a catalyst. The scaling relations thus allow reaction energetics to be estimated based on a relatively smaller number of DFT calculations, which significantly improves the computational tractability. Finally, we use microkinetic modeling to examine reaction hierarchy in great detail. Our results are in excellent agreement with experimental results from the literature. This work serves as an important step for developing models for liquid phase heterogeneous catalysis, and ultimately, to design catalysts for such conditions.

Copyright © American Institute of Chemical Engineers. All rights reserved.

570059 DNA Adsorption on Graphene: DFT Modeling Approach

Graphene, a single layer of carbon in a hexagonal lattice, provides lots of potential for a variety of applications in science and technology due to its excellent electronic properties as well as physical and chemical properties. Pristine graphene is used in this study. Deoxyribonucleic acid (DNA), a complex polymer present in nearly all life forms, contains all the information necessary to build and maintain an organism. Single-stranded DNA is essentially a polymer comprised of four unique deoxyribonucleotide monomers. When deoxyribonucleotide monomers are placed on graphene, they interact creating an energy favorable configuration for the deoxyribonucleotides. The purpose of this study is to investigate the deoxyribonucleotide adsorption on pristine graphene. First, we obtained the geometry-optimized structures of the four deoxyribonucleotides using Density Functional Theory (DFT) calculations. We then obtained the geometry-optimized conformations of the deoxyribonucleosides on pristine graphene using DFT calculations. From those conformations, we calculated the binding energy of the deoxyribonucleotides with graphene. Furthermore, we obtained the potential energy curve of the system as a function of intermolecular distance between deoxyribonucleotides and pristine graphene using DFT single-point energy calculations. From these results, we develop a force field to describe DNA exploring a variety of conformations on pristine graphene at various temperature conditions.

Copyright © American Institute of Chemical Engineers. All rights reserved.

570071 Multiscale Modeling of the UV-Sensitive Spiropyran/Merocyanine Micelles

Multiscale Modeling of the UV- Sensitive Spiropyran/Merocyanine Micelles

Mackenzie Mallard, Seung Min Lee, Connor Callaway, and Seung Soon Jang*

Computational NanoBio Technology Laboratory, School of Materials Science and Engineering, Georgia Institute of Technology, 771 Ferst Drive, Atlanta, GA 30332

Reaction design optimization is an innovative field that drastically improves facets of polymer manufacturing, pharmaceutical production, and related industries. More specifically, the potential that multicompartment micelles (MCM) have to be efficient nanoreactors has caused MCM research to gain popularity. Not only can micelles be used as immobilized molecular catalysts, but they also possess high reactivity and selectivity. Due to the complexity of the multicompartment micelle structure, dissipative particle dynamics (DPD) simulations have been used to model the behavior of spiropyran and reactants in micelles. Spiropyran is a photochromic organic compound used in medical applications. The introduction of UV to this photochromic material causes a change of hydrophobicity which consequently drives a change in the micelle morphology. In our present work, we apply miscibility analyses in order to prepare a coarse-grained model of a tri-polymer micelle with spiropyran. The micelle will consist of a hydrophilic block, a spiropyran block, and a hydrophobic block. The strength of the interaction between a pair of polymer blocks can be measured by the χ-parameter whose value is also dependant on process conditions such as temperature and composition of a solution. In order to run DPD simulations, χ-parameter calculations must be obtained to ensure an accurate representation of the conversion of spiropyran to merocyanine within the micelle. The change in hydrophobicity that results from the conversion of spiropyran to merocyanine causes the change of micelle structure. Using coarse-grained modeling, we investigate the self-aggregation of a spiropyran micelle and the changes in morphology due to the photochromic nature of spiropyran/merocyanine. Because these micelles possess the potential for multi-step catalysis, the effects of the introduction of reactants to the spiropyran micelle are studied, as well as how reactants behave when spiropyran is converted to merocyanine.

Copyright © American Institute of Chemical Engineers. All rights reserved.

570443 Artificial Neural Networks for Accurate Prediction and Analysis of Perovskite Bandgaps

Perovskites are a unique class of semiconductors with long carrier diffusion lengths which have shown remarkable promise for photovoltaic applications as their solar conversion efficiencies increased rapidly by nearly 20% over the past decade. Here we evaluate the efficacy of artificial neural networks for the analysis and prediction of 3D perovskite band gaps; which is the key parameter used in solar cell design. The set of input parameters for our artificial neural network includes descriptors such as s, p, d, and f orbital radii, electronegativity, octahedral factor, and formation energy. Furthermore, we apply a recursive feature correlation filter to eliminate features which highly correlate with other features in our model as well as features which have a low contribution with regards with regards to predicting bandgap. We find that splitting the dataset between training, testing, and validation sets in a 70:15:15 ratio provides a highly reliable model that yields a mean square error of less than 0.25 eV. We then validate the capability of artificial neural network by testing its predictability for bandgaps based on unseen inputs.

Copyright © American Institute of Chemical Engineers. All rights reserved.

571098 First-Principles Study on Developing $Li_{10-X}SnP_2S_{12-X}Cl_X$ Electrolyte with Outstanding Li-Ion Conductivity for High-Voltage Solid State Batteries

First-Principles Study on Developing $Li_{10-x}SnP_2S_{12-x}Cl_x$ Electrolyte with Outstanding Li-ion Conductivity for High-Voltage Solid State Batteries

Kyungju Nam[1], Jeemin Hwang[2], Byungchan Han[1,2,*]

[1] Department of Vehicle Convergence Engineering, Yonsei University, Seoul 03722, Republic of Korea

[2] Department of Chemical and Biomolecular Engineering, Yonsei University, Seoul 03722, Republic of Korea

Using first-principles density functional theory calculations (DFT) and ab-initio molecular dynamics (AIMD) simulations, we suggest $Li_{10-x}SnP_2S_{12-x}Cl_x$ (LSPSC) for the first time as solid electrolyte with supreme ionic conductivity: the substitution of Sn^{4+} for Ge^{4+} and the partial substitution of Cl^- for S^{2-} to a $Li_{10}GeP_2S_{12}$ (LGPS) electrolyte which is well-known sulfide electrolyte. The crystal structure of the LSPSC has a similar symmetry as LGPS, 1 dimensional ion diffusion channels, but further improves on the Li-ion conductivity at room temperature. The main reason of the improvement is chlorine ions that enlarge diffusion channels; however, excess amount of Cl doping degrades the transport of Li ion due to attractive interaction of Cl with Li. The key is to find an optimum halide doping concentration that can maximize the Li-ion conductivity. Calculated density of states (DOS) for LSPSC shows electrochemical instability with approximately 3eV band gap, an insufficient value for the electrolyte material. However, the halides near anode react with Li metal and form LiCl thin layer which demonstrates high bandgap (about 8eV). The computational design concept for a sulfide solid electrolyte with halide doping is a promising for high-voltage solid state batteries.

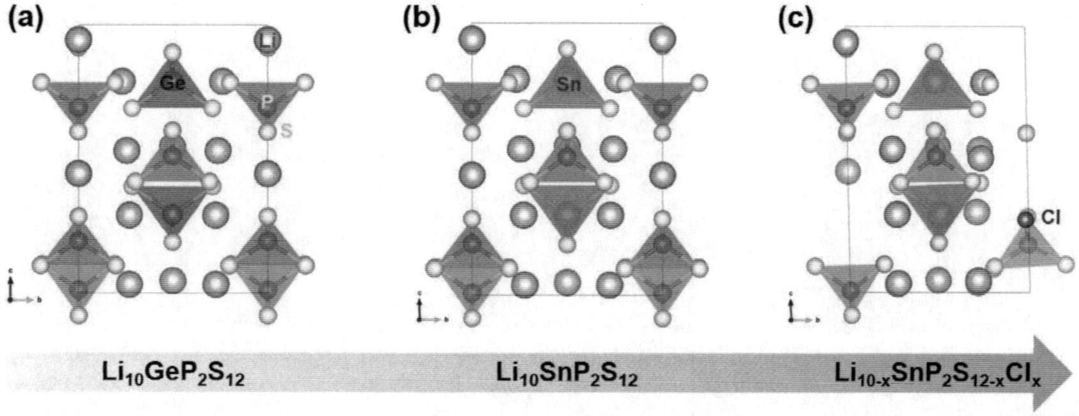

Figure. The model systems of (a) pristine LGPS, (b) LSPS and (c) LSPSC, the partial substitutional doping of Cl to S for LSPS. All model systems have the BCC structure with 1 dimensional diffusion path for Li ions.

Copyright © American Institute of Chemical Engineers. All rights reserved.

571104 First-Principles Study on Unexpectedly Large Lattice Expansion of PdH_x Nanoparticles Generated in Situ in TEM Graphene Liquid Cell

Palladium hydride (PdH_x) nanoparticles(NP) are synthesized in situ in graphene liquid cell under transmission electron microscopy (TEM). During the synthesis, radiolysis of water by the electron beam from TEM produces hydrogen radicals that interact with Pd ions to form PdH_x. In the course of PdH_x NP growth, unexpectedly large lattice expansion is observed, which greatly deviates from what is previously studied from bulk PdH_x. Here, we demonstrate the reason for the unexpected lattice expansion using first-principles density functional theory (DFT) calculations. Firstly, thermodynamic stabilities of cuboctaheral and icosahedral PdH_x NP at different compositions with various hydrogen positions are studied via Energy Convex Hull diagram. Then, hydrogen behaviors on (111) and (100) Pd surfaces are analyzed by calculating adsorption energies in different conditions. Lastly, Nudged Elastic Band (NEB) calculation is utilized to investigate the activation barriers for hydrogen insertions through particular directions. Our results strongly suggest that metastability of PdH_x NP structures and asymmetric hydrogen adsorption are the main keys responsible for the lattice expansion.

Copyright © American Institute of Chemical Engineers. All rights reserved.

572304 Structure and Gas Transport Properties of Organic Mixed Matrices: Polymers of Intrinsic Microporosity with Organic Molecules of Intrinsic Microporosity

Polymers of intrinsic microporosity (PIMs) are microporous materials that have attracted attention for their superior solution processability and microporosity, extremely high free volume and surface areas, which make them good candidates for industrial gas storage and separation. However, the physical aging of PIM membranes can lead to a significant reduction in permeability, solubility, and diffusivity over time. Mixed matrix membranes, by incorporation of an inorganic, metal-organic or organic filler into PIM based polymer matrices, have the potential to improve selectivity and reduce aging effect while maintaining permeability. One of the potential organic filler is the large, rigid organic molecules of intrinsic microporosity (OMIMs) which are known to form porous materials due to their inability to pack efficiently in the solid state.

In the present work, organic mixed matrices (MMs) obtained by the addition of different types of OMIMs into PIM-1 were investigated using atomistic molecular simulations. The structure generation of amorphous Organic MMs was achieved by a simulated polymerization algorithm implemented in the open-source software PySIMM and Polymatic. Structural analysis of six different PIM-1/OMIM mixed matrices, with filler wt% in the range of 7-13%, indicated that using OMIMs as fillers can maintain or even increase the free volume and surface area of the organic MM compared to pure PIM-1. Moreover, pore size distributions of the PIM-1/OMIM MM showed the formation of larger pores with OMIMs frustrating the packing. To compare the gas transport properties of these organic MMs with pure PIM-1, grand canonical Monte Carlo was used to obtain the adsorption isotherms of CO_2, N_2, O_2, and CH_4 in these materials. Plasticization behavior of one of the PIM-1/OMIM mixed matrices was also studied using a combined Monte Carlo and molecular dynamics simulation.

Copyright © American Institute of Chemical Engineers. All rights reserved.

572632 Transport Mechanisms of Polar Solutes in a Cross-Linked Hii Phase Lyotropic Liquid Crystal Membrane

Cross-linked inverted hexagonal phase lyotropic liquid crystal (LLC) membranes possess pores of uniform size with a complex chemical topology that make them promising for selective separations. We characterized transport of water, sodium ions and 20 small polar solutes within the pores of an LLC membrane using atomistic molecular simulations. Due to the inhomogeneous pore compostion, we observed radially dependent transport mechanisms that are a strong function of solute chemical functionality. In general, all solutes perform intermittent hops between lengthy periods of entrapment. Three different trapping mechanisms are responsible for this behavior. First, solutes that drift out of the pore can become entangled among the dense monomer tails. Second, solutes can donate hydrogen bonds to the monomer head groups. Third, solutes can coordinate with sodium counter ions. The degree to which a solute is affected by each mechanism is dependent on the chemical functionality of the solute. We have used our detailed mechanistic understanding in order to construct a stochastic model which we can use to project the long term behavior of the solutes and ultimately predict selectivity. Using the insights developed in this study, we can begin to think about how to redesign existing LLC membranes in order to perform solute-specific separations.

Copyright © American Institute of Chemical Engineers. All rights reserved.

573201 Towards Rational Design of Chemical Modulators to Improve Plant Drought Tolerance

Drought is a global environmental challenge that severely affects crop yields, and in turn threatens food and biofuel production. Abscisic acid (ABA) is a hormone molecule produced in plants under water deficiency, which eventually leads to guard cell closure to reduce transpiration water loss. ABA is recognized by a family of receptor proteins that have 14 members (PYR1 and PYL1-13) in *Arabidopsis thaliana*. Notably, PYL4-PYL13 are monomers in solution, while PYR1 and PYL1-3 form a homodimer. Upon ABA binding to the structurally conserved, water-filled binding site, ABA receptors undergo a pronounced conformational change to activate, which triggers the signaling cascade. ABA receptors are key molecular targets for chemical manipulation of plant water use efficiency. Due to ABA's poor stability and functional redundancy of receptors, considerable research efforts have been expended on agrochemical discovery and ABA receptor engineering in order to achieve selective activity control of ABA receptors.

While ABA signaling mechanism has been intensely investigated, several aspects on activation of ABA receptors and their downregulation by post-translational modifications (PTMs) remain elusive. Specifically, the monomeric receptors exhibit ABA-independent constitutive activity, while the dimeric receptors are ABA-dependent and have two orders of magnitude lower ABA affinity. Also, the dissociation of the dimeric receptors, facilitated by the presence of ABA, is required for full activation. In addition, ABA receptors are subjected to various types of PTMs, which can render current agrochemicals ineffective. The unclear molecular origins of these observations substantially hamper the rational design of ABA agonists with high stability, affinity and selectivity.

In this work, we have utilized unbiased molecular dynamics (MD) simulations to elucidate the complete molecular basis of ABA-mediated activation of three subtype receptors (dimeric PYL2 and monomeric PYL5, PYL10) and a modified PYL5 through tyrosine nitration. We have performed ~500 μs (aggregate) MD simulations to capture ABA binding and subsequent receptor activation in atomic details. We then apply the cutting-edge Markov state models (MSM) methodology to fully map out the protein-ligand binding process and the activation pathways of the three ABA receptors, along with quantitative thermodynamic and kinetic characterizations. Energetic landscapes reveal that ABA binding is necessary but insufficient for full receptor activation. ABA must surmount a large energy barrier to bind the receptors and the major barrier appears to be associated with substantial dewetting of receptor during ABA binding. Our results explain the molecular mechanisms of the ABA-independent activation for the monomeric PYL10 receptor and the ABA-dependent activation for the dimeric PYL2 receptor. We show that the tyrosine nitration of PYL5 significantly alters the binding site, thereby preventing ABA perception.

To study the effects of ABA binding on the dissociation of the dimeric PYL2 receptor, we have employed replica exchange umbrella sampling (REUS) simulations to accurately calculate the standard PYL2 association free energy. Coupled with multistate Bennett acceptance ratio (MBAR) method, we determine that the protein-protein binding free energies for the apo-apo, the holo-apo, the holo-holo PYL2 receptors are -14.118 kcal/mol, -9.661 kcal/mol and -9.241 kcal/mol. These results suggest that the binding of ABA causes ~5 kcal/mol difference in PYL2 association free energy, thereby destabilizing the PYL2 complex.

Finally, we have investigated the solvation thermodynamics of ABA receptors via MD simulations and inhomogeneous solvation theory. By comparing the water thermodynamics at the binding site in the presence and the absence of ABA, we obtain energetic insights into the role of water molecules on ABA binding affinity and selectivity. We demonstrate that exploiting water thermodynamics of ABA receptors can be a promising strategy for modifying and designing ABA agonists.

Reference: Shukla S.[+], Zhao C.[+], & Shukla D. (2019). Dewetting Controls Plant Hormone Perception and Initiation of Drought Resistance Signaling. *Structure*. 27, 692-702.

Copyright © American Institute of Chemical Engineers. All rights reserved.

573208 Rosette Nanotubes for Applications in Water Filtration, Biosensing and Drug Delivery: A Molecule Dynamics Study

Biocompatible supramolecular nanostructures, such as the rosette nanotubes (RNTs), are formed via self-assembly of building blocks of Watson-Crick DNA-inspired guanine-cytosine (G C) motifs. Hydrogen bonding between the individual motifs, much like the double helix of DNA, lead them to assemble into rings, called rosettes.[1,2] The rosettes then self-assemble into nanotubes, where the rings either stack or assemble into helical coils through a combination of π-π interactions between the rings and hydrophobic effects. Due to its biocompatibility, RNTs have been used for drug delivery and biological applications, ie. enhancing cell growth in bones by encapsulating dexamethasone.[3] In addition, other potential applications of RNTs can include nanopore-based sensors and water filtration membranes. Here we report on our current studies that involve (1) modeling the interactions of RNTs with cyclodextrins and water, and (2) developing a Martini-based[4] coarse-grained model for the RNTs.

1. Fenniri, B. L. Deng, and A. E. Ribbe. *J. Am. Chem. Soc.* 2002, **124**, 11064-11072.
2. Fenniri, B. L. Deng, A. E. Ribbe, K. Hallenga, J. Jacob, and P. Thiyagarajan, *Proc. Natl. Acad. Sci. USA* 2002, **99**, 6487-6492.
3. Borzsonyi, R. Johnson, A. Myles, J. Cho, T. Yamazaki, R. Beingessner, A. Kovalenko, and H. Fenniri. *Chem. Commun.*, 2010, **46**, 6527-6529.
4. Uusitalo, H. Ingólfsson, P. Akhshi, D. Tieleman, and S. Marrink. *J. Chem. Theory Comput.*, 2015, **11**, 3932–3945.

Copyright © American Institute of Chemical Engineers. All rights reserved.

573541 Investigating the Core-Shell Structure of Soot Using Molecular Modelling Methods

Soot particle structure: Insights from molecular dynamics and Monte Carlo simulations

Kimberly Bowal[a], Jacob W. Martin[a,b], Peter Grancic[c], Markus Kraft[a,b,d]

[a]Department of Chemical Engineering and Biotechnology, University of Cambridge, West Site Philippa Fawcett Drive, Cambridge, CB3 0AS UK

[b]Cambridge Centre for Advanced Research and Education in Singapore (CARES), CREATE Tower, 1 Create Way, 138602 Singapore

[c]Institute of Soil Research, University of Natural Resources and Life Sciences, Peter-Jordan-Straße 82, 1190 Vienna

[d]School of Chemical and Biomedical Engineering, Nanyang Technological University, 62 Nanyang Drive, 637459 Singapore

Incomplete combustion produces carbonaceous particulate matter, known as soot, which negatively affects combustion devices, human health and the environment. The development of soot prediction models and mitigation strategies requires an accurate knowledge of soot particle formation and structure. Electron microscopy has revealed that mature soot particles often have a core-shell structure, in which the molecules near the centre of the particle differ in size from those in the surface layers [1]. However, it is unknown what interactions cause this partitioning and to what extent it is present in nascent particles.

In this work, the behaviour of heterogeneous polycyclic aromatic hydrocarbon (PAH) clusters was investigated using two independent molecular modelling methods to understand the structure of nascent soot particles. Clusters of up to 100 molecules containing combinations of the different sized PAHs circumcoronene, coronene, ovalene, or pyrene were evaluated. Replica exchange molecular dynamics (REMD) simulations sampled many configurations at high and low temperatures to determine stable low energy structures [2]. These results were compared with a novel Sphere Encapsulated Monte Carlo (SEMC) method, developed here to extend basin-hopping minimisation methods for use with planar aromatic molecules.

Both the REMD and SEMC simulations showed that the most stable cluster structures consist of parallel columns of stacked PAHs in a core-shell structure, in which the larger PAHs are located in the core, and the smaller PAHs are found in a surrounding shell, as seen for a representative cluster in Figure 1. This is an inverse partitioning compared to that seen in mature soot particles, proposing a unique structure for nascent soot particles and suggesting that the molecular organisation determined by intermolecular interactions is not responsible for the core-shell structure of mature soot. These results present the first simulations of soot-sized heterogeneous PAH clusters and can help further understanding of soot formation and growth processes.

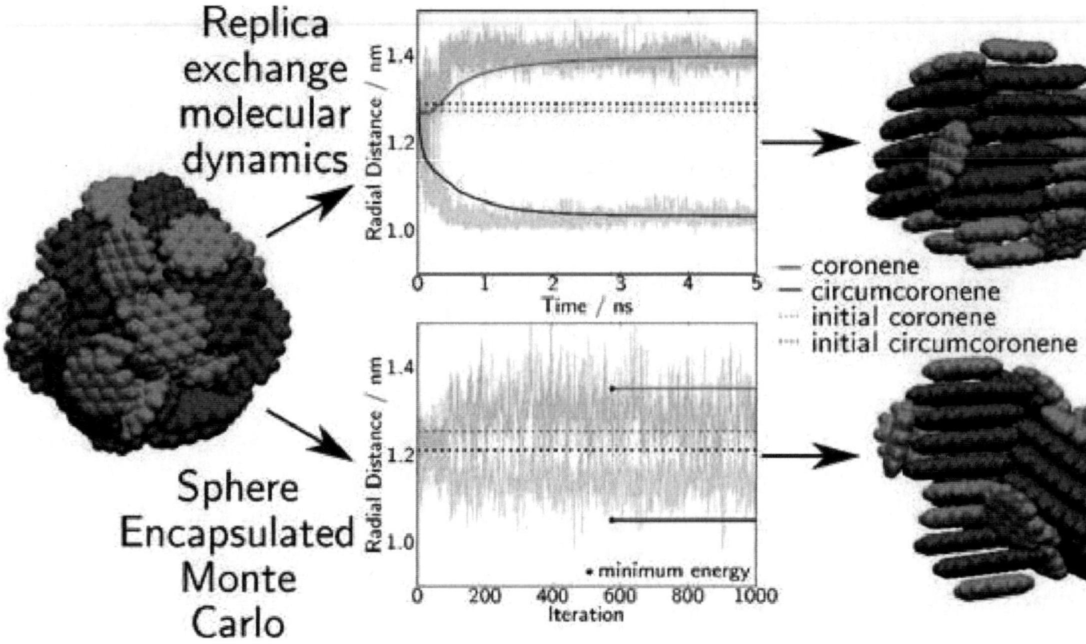

Figure 1: Radial distances of each molecule type shown for a cluster containing 16 coronene molecules (coloured red) and 16 circumcoronene molecules (coloured blue) using replica exchange molecular dynamics simulations and the Sphere Encapsulated Monte Carlo method. Initial and final cluster snapshots are also shown.

[1] T. Ishiguro, Y. Takatori, and K. Akihama. *Combust. Flame,* (1997), **108**, 231–234.

[2] K. Bowal, J.W. Martin, M. Kraft. *Carbon*, (2019), **143,** 247-256.

Copyright © American Institute of Chemical Engineers. All rights reserved.

574348 Atomistic Simulation Studies of Artificial Water Channels in Membranes

Peptide appended pillar[5]arene (PAP) is an artificial water channel resembling biological water channel proteins, which has shown promise for designing bioinspired water purification systems. Insertion of PAP channels into a membrane matrix is required for applications in separation devices, which unavoidably results in membrane perturbations due to channel-channel and channel-membrane interactions and may affect the channel performance.

To better understand the nature of interactions between PAP channels and membranes, we have carried out atomistic MD simulation studies of multiple PAP channels embedded in a lipid as well as a polymeric membrane matrix. Specifically, we have characterized clustering propensity of PAP channels and identified several metastable states highlighting differences in channel dynamics in biological (lipid) and synthetic (polymeric) membranes. We will also present results on the functional transport characteristics of PAP channels where we observed wetting/dewetting transitions indicating a gating mechanism.

Copyright © American Institute of Chemical Engineers. All rights reserved.

574359 Deciphering the Effects of a Triple Mutation on the Structure and Dynamics of Human Interferon Alpha 2-a Via Atomistic Simulations

Interferon Alpha 2-A (IFNα2-A) is a cytokine messenger protein and one of the 16 Type-I Interferons, produced by the cells affected by viruses, bacteria or cancer—to warn nearby healthy cells to uplift their defense against possible infections. IFNα2-A shows anti-viral and anti-proliferative activity and it is used to treat chronic hepatitis B, C, and cancers. All Type-I IFNs signal through the same two receptors: IFNAR1 and IFNAR2; and depending on their binding—the responses vary, making them multifunctional. As the multifunctionality of IFNα2-A causes various side-effects, many mutants such as the triple mutant HEQ (His57Ala, Glu58Ala, and Gln61Ala) were produced intending to minimize the side-effects. HEQ was found to show altered anti-viral and anti-proliferative potencies when compared with wild-type, yet, a proper understanding of the effects of the triple mutation on the structure and dynamics was lacking.

This work presents the results of an atomistic molecular dynamics study carried out to compare various structural and dynamic properties of wild-type IFNα2-A, and its triple mutant, HEQ. The initial structural analysis comparing the two systems showed that the mutation has a minimal effect on the overall structural fold of the protein, whereas the B factor calculations—along with the per residue RMSD and contacts analysis of the IFNAR1 binding site residues—have indicated changes in the local structure and dynamics of the binding site. Further, the dynamic cross-correlation analysis of the two systems has indicated a vivid disparity, showing more negatively correlated residue motions for the mutant and more positive correlations for the wild-type.

These results are expected to provide insights on how the mutation affects the binding of the protein to its receptor subunits when forming the final signaling complex.

Copyright © American Institute of Chemical Engineers. All rights reserved.

574395 Molecular Dynamics Simulations of New Ionic Liquids with Crystal-like Structures As Highly Selective Membrane Materials for CO_2 Separation

For over a decade, ionic liquids (ILs) and polymerized ILs have been investigated as advanced materials for CO_2 capture and separation applications.[1] Herein, we are exploring a new approach for achieving high performance (high selectivity, high diffusivity) IL-based materials for industrial gas separation. By leveraging the strong electrostatic interactions and steric constraints present within different IL formulations, we are designing highly-ordered, liquid-crystal like structures that maximize gas diffusivity while simultaneously preserving selectivity. This is a combined computational and experimental investigation of unique IL architectures, including $C_{14}H_{16}N_4^{2+}$; $C_{18}H_{21}N_6^{3+}$; $C_{22}H_{26}N_8^{4+}$; $C_{10}H_6O_6S_2^{2-}$; $C_{12}H_6F_6N_2O_8S_4^{2-}$; $C_6H_3O_9S_3^{3-}$ combined within in a methanol or N-methyl-2-pyrrolidone (NMP) solvent.

In our work, we are using molecular dynamics methods to provide information about how the cations and anions interact with each other at a molecular level. Our simulation techniques are being used to capture information about structural correlation functions, molecular diffusivity, as well as an explicit accounting of the balance of van der Waals and electrostatic interactions in the system. Also, due to the porous nature of our materials, we also provide quantitative information about the fractional free volume (FFV), surface area, and pore size distribution.[2] These properties can provide useful screening information for predicting gas adsorption/separation performance, and this is being verified by directly simulating gas adsorption isotherms with grand canonical Monte Carlo (GCMC) simulations. Our simulations are being rigorously benchmarked against the on-going experimental investigations, and this will provide essential information for future experimental and computational design of highly selective membrane materials.

REFERENCES

1. Li, P.; Paul, D.; Chung, T.-S., High performance membranes based on ionic liquid polymers for CO_2 separation from the flue gas. *Green Chem.* **2012,** *14*, 1052-1063.
2. Abedini, A.; Crabtree, E.; Bara, J. E.; Turner, C. H., Molecular Simulation of Ionic Polyimides and Composites with Ionic Liquids as Gas-Separation Membranes. *Langmuir* **2017,** *33*, 11377-11389.

Copyright © American Institute of Chemical Engineers. All rights reserved.

Prediction of Rheological Properties of LDPE Produced in High-Pressure Tubular Reactors.

Maira L. Dietrich[1], Adriana Brandolin,[1, 2] Claudia Sarmoria[1, 2] and Mariano Asteasuain[1, 2]

[1] *Planta Piloto de Ingeniería Química (PLAPIQUI), Universidad Nacional del Sur-CONICET, Bahía Blanca, Buenos Aires, Argentina.*

[2] *Dpto. de Ingeniería Química, Universidad Nacional del Sur (UNS), Bahía Blanca, Buenos Aires, Argentina.*

Abstract

This work presents a high-fidelity model of a high-pressure tubular reactor for the production of low-density polyethylene able to predict rheological properties of the polymer. A previous model developed by the authors,[1] which calculated molecular properties such as the molecular weight distribution, the joint molecular weight-long chain branching distribution and molecular weight-short chain branching distribution, and average molecular properties, was updated to include the prediction of rheological properties such as the reptation modulus, the storage and loss moduli and the complex viscosity. The latter is used as input to compute the polymer melt index. The model allows analyzing the influence of operating conditions on molecular structure and on the rheology of the polymer melt. This knowledge allows producers to design the material molecular structure in order to meet a particular processing behavior.

Introduction

Distributed molecular properties of polyolefins, such as molecular weight distribution (MWD), long-chain branching distribution (LCBD) and short-chain branching distribution (SCBD) are strongly related to the end-use properties of the polymer. Depending on the molecular characteristics, different grades of a same polymer may present different rheological and mechanical properties, leading to a distinct processability behavior. Predicting viscoelastic response of polymer melts from the knowledge of molecular structure is an important point still in continuous development in polymer science. Models with these capabilities can be used to predict processing performance of the melt based on the polymer chain microstructure properties.

One common approach to modeling the polymer melt rheological response is based on the reptation concepts proposed by de Gennes[2] and subsequently developed by Doi and Edwards[3]. They proposed a mathematical model to predict the rheological behavior of linear polymer melts in terms of the MWD. According to their theory, a particular polymer chain, called test chain, is surrounded by other entangled polymer chains that act as a tube made of fixed obstacles restricting its motions. The motion of the chain along this tube is approximated as curvilinear reptation motion of its repetitive units. In this basic theory relaxation of the chains forming the tube is supposed to have no influence on the test chain. The original model by Doi and Edwards[3] has shown several limitations in modeling viscoelastic properties of polymers, motivating the introduction of different modifications to this this theory. In addition to pure reptation by curvilinear diffusion of the test chain, there are other relaxation mechanisms such as tube length fluctuations and constraint release mechanisms that play an important role.[4, 5]

For these reasons the simple reptation model of Doi and Edwards[3] is restricted only to qualitative predictions, even in the simplest case of purely linear monodisperse polymer melts. The theory of double reptation of Tsenoglou[6] and Des Cloizeaux[7] has also been used to predict polymer viscoelasticity properties from MWD knowledge. In this theory, if two polymer chains (of different or same molecular weight) exert stress on each other at an initial time, this stress will hold at a later time if neither of them slips through the stress point. The stress relaxation function is calculated as the combination of the probabilities that each polymer slip through the stress point between the initial and an arbitrary time.[8] To complete this theory, the probability for each polymer chain has to be calculated. For this purpose these authors proposed the time-dependent diffusion model, which also describes a chain relaxing in a tube made of fixed obstacles, but consider a time dependent diffusion coefficient.

A limitation of the above described models comes from the fact that synthetic polymers are usually polydisperse. Although the flow properties of entangled melts depend strongly on the MWD of the polymer, most of the models available in literature have only been tested for monodisperse or binary mixtures of polymers of the same chemical nature but different molecular weight. The dependence of the viscoelastic properties on the molecular weight structure of a mixture determines the mixing behavior, which is reflected in terms of a mixing rule.

Some mixing rules have been proposed that weights the contribution of various chains lengths and the interaction between them on the relaxation modulus. Doi and Edwards[3] proposed a linear mixing rule where the relaxation modulus of the mixture is calculated as the linear superposition of the relaxation function of each of the components multiplied by its molecular weight. Later, Des Cloizeaux[7] and Tsenoglou[6] developed a quadratic mixing rule where the squared root of the relaxation modulus is obtained summing up the squared roots of the weighted relaxation functions of the components. The linear mixing rule can be derived theoretically from the reptation model, while the quadratic mixing rule was derived using the double reptation concept. Besides, different relaxation functions have been proposed in the literature, for example the Doi and Edwards function,[3] the Tuminello step function,[9] the single exponential function[10] and the des Cloiseaux function.[11] These mixing rules are inadequate to describe the relaxation modulus of a polymer in some cases, so phenomenological mixing rules have been derived to fit experimental data have also been proposed.[12]

In the present work, the time-dependent diffusion model of des Cloizeaux[11] was employed to qualitatively predict the viscoelastic behavior of low density polyethylene (LDPE) produced in a high-pressure tubular reactor using its MWD as input data. Employing an appropriate mixing rule and the relaxation function proposed by the time-dependent diffusion model, the relaxation modulus is estimated. This modulus, in turn, is used to estimate the storage and loss moduli using the Schwarzl[13] relations and subsequently the complex viscosity of different grades of LDPE. This approach has been previously applied by other authors to predict the viscoelastic behavior of several linear polyolefins showing qualitatively good results.[14, 15] Besides, it has been also proposed as a method to detect very low levels of LCB by analyzing discrepancies in the terminal region of the storage and loss moduli curves.[16]

Finally, the high-pressure tubular reactor model also predicts the melt index (MI) by means of a model that employs the complex viscosity curve as input data. The MI is one of the most common rheological parameters used by polymer producers for quality control. Frequently, the MI is continuously measured during production to control that the material presents rheological characteristics appropriate for its correct further processing. For this reason, an integrated model that can relate the reactor operation conditions, the molecular properties and finally rheological parameters results of a great importance.

Mathematical model

The double reptation theory proposed by Des Cloizeaux[8],[11] and Tsenoglou[10] predicts the relaxation modulus $G(t)$ of an entangled linear polymer from its MWD (expressed as weight fraction $w(M)$ of chains having molecular weight M) through the following mixing rule:

$$G(t) = G_N^0 \left(\int_{Me}^{\infty} (F_{MD}(t,M))^{1/\beta} w(M) dM \right)^{\beta} \tag{1}$$

In this equation, G_N^0 is the plateau modulus and M_e is the molecular weight between entanglements. In the double reptation theory the exponent β is set to 2, but in the present work we used the proposed modified value of 2.25 suggested by other authors.[14, 15, 16] Variable F_{MD} is the relaxation function corresponding to a monodisperse polymer of molecular weight M. In this work, the time-dependent diffusion reptation model developed by des Cloizeaux[11] is employed:

$$F_{MD}(t,M) = \frac{8}{\pi^2} \sum_{podd} \frac{1}{p^2} \exp\left(-p^2 U(t,M)\right) \tag{2}$$

$$U(t,M) = \frac{t}{\tau_{rept}(M)} + \frac{M^*}{M} f\left(\frac{Mt}{M^* \tau_{rept}(M)}\right) \tag{3}$$

f is a function that can be approximated as:

$$f(x) = -x + x^{0.5}\left(x + (\pi x)^{0.5} + \pi\right)^{0.5}$$
$$x = \left(\frac{Mt}{M^* \tau_{rept}(M)}\right) \tag{4}$$

The reptation time τ_{rept} is related to M by:

$$\tau_{rept}(M) = KM^3 \tag{5}$$

M^* and K are both adjustable parameters.

The above model includes four unknown parameters: G_N^0, M_e, M^* and K. The first of them acts as a scaling factor in the $G(t)$ formula of Eq.(1). Experimental values of this parameter are found in literature for different grades of polyethylene[17], from which the value of M_e can be obtained as $G_N^0 = \frac{\rho RT}{M_e}$. Here, ρ is the polymer density, R the ideal gas constant and T the temperature of the polymer melt. In the present work, the values of the four parameters where taken from Van Ruymbeke et al.[15] who obtained these values for a slightly branched HDPE, as: $G_N^0(Pa) = 2.6E^6$, $M^*(g/mol) = 70000$ and $K(s(mol/g)^3) = 1.4E^{-17}$.

The storage moduli $G'(t)$ and the loss moduli $G''(t)$ are calculated as a function of $G(t)$ using the Schwarzl[13] approximations:

$$\begin{aligned} G'(w) = {} & G(t) - 0.142(G(4t) - G(8t)) + 0.717(G(2t) - G(4t)) + 0.046(G(t) - G(2t)) \\ & + 0.099(G(t/2) - G(t)) + 0.103(G(t/4) - G(t/2)) + 0.001(G(t/8) - G(t/4)) \\ & + 0.00716(G(t/16) - G(t/8)) + 0.000451(G(t/64) - G(t/32)) \end{aligned} \tag{6}$$

$$\begin{aligned} G''(w) = {} & -0.470(G(2t) - G(4t)) + 1.674(G(t) - G(2t)) + 0.198(G(t/2) - G(t)) \\ & + 0.620(G(t/4) - G(t/2)) + 0.012(G(t/8) - G(t/4)) \\ & + 0.172(G(t/16) - G(t/8)) + 0.043(G(t/64) - G(t/32)) \\ & + 0.0108(G(t/256) - G(t/128)) \end{aligned} \tag{7}$$

From these equations the complex viscosity $\eta^*(w)$ can be approximated as:

$$\eta^*(w) = \frac{1}{w}\sqrt{G'(w)^2 + G''(w)^2} \tag{8}$$

To obtain the MI value, which condensates the rheological information of the polymer melt, a physical description of the flow in the melt-indexer is proposed. The model considers that the sum of the pressure drops along the different sections of the melt-indexer apparatus, which are the barrel (P_b), the capillary (P_c) and the entrance region of the capillary (P_e), is equal to the total applied pressure:

$$P_b + P_e + P_c = {Mg}/{\pi R_p^2} \tag{9}$$

In this equation, g is the gravity constant, M the applied mass, and R_p is the piston radius. All the melt-indexer geometrical parameters (radius and lengths) as well as the applied mass are specified by the ASTM D-1238 norm. Using the definition of volumetric flow in a tube, the Weissenberg-Rabinowitsch equation can be derived trough mathematical manipulation.[18] From this equation, for fluids that follow the power law model for the shear viscosity curve $\eta(\dot{\gamma})$, a relation can be obtained between the volumetric flow Q and the slope n. In this way, equations for Q in the barrel and in the capillary are obtained:

$$Q = \frac{\pi R_b^3 \dot{\gamma}_{R_b}}{4}\left(\frac{4n}{3n+1}\right) \tag{10}$$

$$Q = \frac{\pi R_c^3 \dot{\gamma}_{R_c}}{4}\left(\frac{4n}{3n+1}\right) \tag{11}$$

Additionally, from the momentum balance in the barrel and capillary it is obtained:

$$\dot{\gamma}_{R_b}\eta(\dot{\gamma}_{R_b}) = P_b R_b / 2L_b \tag{12}$$

$$\dot{\gamma}_{R_c}\eta(\dot{\gamma}_{R_c}) = P_c R_c / 2L_c \tag{13}$$

In these equations, R_c and R_b are the capillary and barrel radius, L_c and L_b the capillary and barrel length, and $\dot{\gamma}_{R_c}$ and $\dot{\gamma}_{R_b}$ the shear rates at the capillary and barrel wall, respectively. $\eta(\dot{\gamma})$ is the shear viscosity, which taking into account the Cox-Merz rule at low frequencies, is approximated using the complex viscosity by: $\eta(\dot{\gamma}) \approx \eta^*(w)$. The pressure drop at the capillary entrance is estimated through the Cogswell[19] equation:

$$P_e = \frac{4\sqrt{2}\dot{\gamma}_a}{3(n+1)}\sqrt{\eta(\dot{\gamma}_a)\eta_e(\dot{\varepsilon})}\left(\frac{4n}{3n+1}\right)^{(1+n/2)} \tag{14}$$

The apparent shear rate $\dot{\gamma}_a$ is related to the shear rate at the capillary wall through:

$$\dot{\gamma}_a = \frac{4}{3+1/n}\dot{\gamma}_{R_c} \tag{15}$$

Additionally, the elongational viscosity is described by:

$$\dot{\varepsilon} = \frac{4\eta(\dot{\gamma}_a)\dot{\gamma}_a}{3(n+1)P_e}\left[\frac{4n}{3n+1}\right]^{n+1} \tag{16}$$

As described in the above equations, to calculate the volumetric flow is necessary to have both the shear and elongational viscosity. Here, we applied the Trouton´s rule[20] by which $\eta_e(\dot{\varepsilon}) \square k\eta(\dot{\gamma})|_{\dot{\varepsilon}=\dot{\gamma}}$. Originally, $k=3$ for Newtonian fluids or polymer melts at low shear rates. However, in the particular case of our LDPE samples, $\dot{\varepsilon}$ presents relatively high values, so this constant was set to 10 following the original guidelines of Trouton[20]. Finally, solving the non-linear system of Eq. (9)-(16), P_b, P_c, P_e, Q, $\dot{\gamma}_{R_b}$, $\dot{\gamma}_{R_c}$, $\dot{\gamma}_a$ and $\dot{\varepsilon}$ are obtained. In these equations, the values of $\eta(\dot{\gamma}_{R_b})$, $\eta(\dot{\gamma}_{R_c})$, $\eta(\dot{\gamma}_a)$ and $\eta_e(\dot{\varepsilon})$ are the computed from the $\eta^*(w)$ curve of the polymer sample. Once Q has been determined, the MI is obtained as:

$$MI\left(\frac{g}{10\min}\right)=Q\left(\frac{m^3}{s}\right)\rho\left(\frac{kg}{m^3}\right)\left(\frac{1000g}{1kg}\right)\left(\frac{60s}{1\min}\right) \tag{17}$$

Results and Discussion

The proposed kinetic and rheological model is used to qualitatively predict the viscoelastic behavior of LDPE produced in high-pressure tubular reactor under different operation conditions. In a previous work developed by the authors[1] we obtained the MWD, MWD-LCBD, MWD-SCBD and average molecular properties such as dispersity (Đ), weight-average molecular weight ($\bar{M}_w$), number-average molecular weight ($\bar{M}_n$) and number of long chain branches / 1000 carbon atoms (LCB/1000C) of several samples of LDPE. In Table 2 of that work, reactor conditions of two operation cases (named Case I and Case II) are summarized. In the present work, we will consider Case I as a base operation case. The polymer obtained under such conditions is characterized by having the majority of chains linear or slightly branched (1-2 branches per chain). Figure 1 shows the MWD obtained for Case I at the reactor exit. Also, the complex viscosity curve at 190°C is shown in Figure 2.

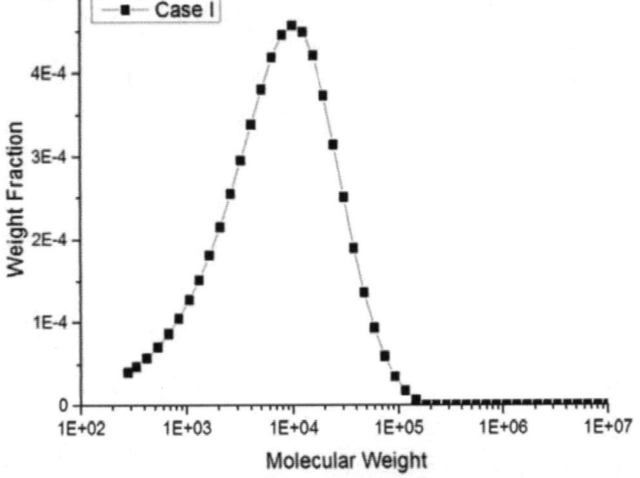

Figure 1. MWD of LDPE produced under operation conditions of Case I of a high pressure tubular reactor.

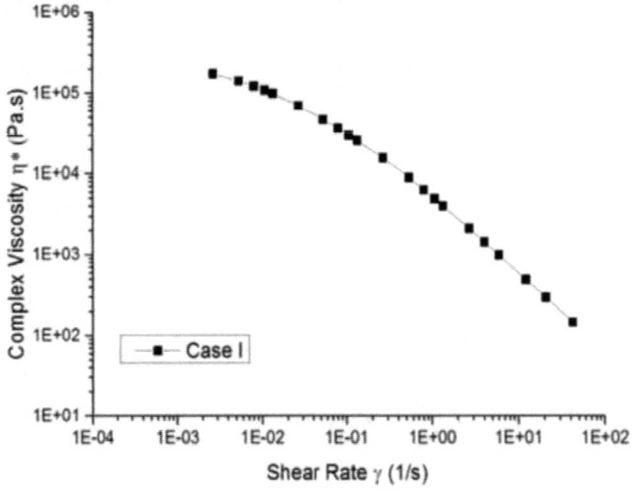

Figure 2. Predicted complex viscosity curve for LDPE produced under operation conditions of Case I of a high pressure tubular reactor.

Figure 3 shows the relation between calculated values of MI obtained by appliyng the matematical model described avobe and the $\bar{M}_w$ of several samples of LDPE. All the samples considered contain about the same content of LCB. The highlighted point in red represents Case I corresponding to the complex viscosity curve in Figure 2. A clear decrease of the MI values with the increase of $\bar{M}_w$ is shown in Figure 3. This behavior can be attributed to the fact that the presence of longer molecules contributes to the generation of more entanglements and therefore grater resistance to flow. As these LDPE samples present similar LCB content, the variation in MI values could be consider entirely due to the molecular weight influence.

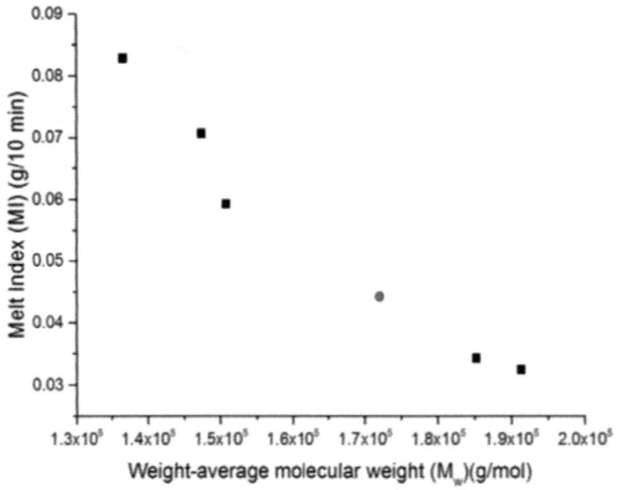

Figure 3. MI vs weight-average molecular weight of several samples of LDPE including Case I highlighted in red.

Conclusion

A mathematical model able to predict molecular and rheological properties of LDPE produced in high-pressure tubular reactors is presented. The model integrates molecular information coming from a kinetic model, such as the molecular weight distribution, long and short chain branching distribution and average molecular properties with a qualitative rheological model. The developed mathematical tool is used to evaluate the effect of different reactor operation conditions on the molecular characteristics of different LDPE samples and subsequent on its rheological properties such as relaxation modulus, complex viscosity and melt index. In this way, we developed a useful tool for polymer producers to characterize and molecular design their polymer production.

References

1. Dietrich, M. L.; Sarmoria, C.; Brandolin, A.; Asteasuain, M. (2019), "LDPE Production in Tubular Reactors: Comprehensive Model for the Prediction of the Joint Molecular Weight-Short (Long) Chain Branching Distributions", *Ind. Eng. Chem. Res.*, 58, 4412-4424.
2. de Gennes, P.-G. (1971), "Reptation of a polymer chain in the presence of fixed obstacles", *J. Chem. Phys.*, 55, 572-579.
3. Doi, M.; Edwards, S. F. (1988), "The Theory of Polymer Dynamics". Vol. 73. Oxford University Press.
4. Leonardi, F.; Majeste, J. C.; Allal, A.; Marin, G. (2000), "Rheological models based on the double reptation mixing rule: The effects of a polydisperse environment", *J Rheol*, 44, 675-692.
5. Des Cloizeaux, J. (1990), "Relaxation of entangled polymers in melts", *Macromolecules*, 23, 3992-4006.
6. Tsenoglou, C. (1987) "Viscoelasticity of binary homopolymer blends. *Abstracts of Papers of the American Chemical Society*, Amer. Chemical Soc. 1155 16TH ST, NW, Washington, DC 20036,194.
7. Des Cloizeaux, J. (1988), "Double reptation vs. simple reptation in polymer melts", *EPL*, 5, 437.
8. Des Cloizeaux, J. (1992), "Relaxation of Entangled and Partially Entangled Polymers in Melts: Time-Dependent Reptation", *Macromolecules*, 25, 835-841.
9. Tuminello, W. H. (1986), "Molecular weight and molecular weight distribution from dynamic measurements of polymer melts", *Polym. Eng. Sci.*, 26, 1339-1347.
10. Tsenoglou, C. (1991), "Molecular weight polydispersity effects on the viscoelasticity of entangled linear polymers", *Macromolecules*, 24, 1762-1767.
11. des Cloizeaux, J. (1990), "Relaxation and viscosity anomaly of melts made of long entangled polymers: time-dependent reptation", *Macromolecules*, 23, 4678-4687.
12. Maier, D.; Eckstein, A.; Friedrich, C.; Honerkamp, J. (1998), "Evaluation of models combining rheological data with the molecular weight distribution", *J. Rheol.*, 42, 1153-1173.
13. Schwarzl, F. R. (1971), "Numerical calculation of storage and loss modulus from stress relaxation data for linear viscoelastic materials", *Rheol. Acta*, 10, 165-173.
14. Pladis, P.; Kanellopoulos, V.; Chatzidoukas, C.; Kiparissides, C. (2008), "Effect of Reaction Conditions and Catalyst Design on the Rheological Properties of Polyolefins Produced in Gas-Phase Olefin Polymerization Reactors", *Macromol. Theory Simul.*, 17, 478-487.
15. Van Ruymbeke, E.; Keunings, R.; Stéphenne, V.; Hagenaars, A.; Bailly, C. (2002), "Evaluation of reptation models for predicting the linear viscoelastic properties of entangled linear polymers", *Macromolecules*, 35, 2689-2699.
16. Van Ruymbeke, E.; Stéphenne, V.; Daoust, D.; Godard, P.; Keunings, R.; Bailly, C. (2005), "A sensitive method to detect very low levels of long chain branching from the molar mass distribution and linear viscoelastic response", *J. Rheol.*, 49, 1503-1520.
17. Ferry, J. D. (1980), "Viscoelastic properties of polymers": John Wiley & Sons.
18. Liu, Y. J.; Qu, J. P.; Cao, X. W. (2005), "Mathematical model of shear rate of polymer melt dynamic extruding through capillary", *J. Appl. Polym. Sci.*, 95, 1056-1061.
19. Cogswell, F. N. (1972), "Converging flow of polymer melts in extrusion dies", *Polym. Eng. Sci.*, 12, 64-73.
20. Trouton, F. T. (1906) "On the coefficient of viscous traction and its relation to that of viscosity. *Proceedings of the Royal Society of London. Series A, Containing Papers of a Mathematical and Physical Character*, 77.

574984 Development of a New Computational Tool to Study Protein Assembly at the Blood-Brain Barrier Interface

Transport of molecules to the brain from the neurovasculature, across the blood-brain barrier (BBB) interface, occurs via two major pathways—paracellular and transcellular. The paracellular pathway is passively regulated by transmembrane proteins that interact with each other to form physical barriers called the tight junctions. The tight junctions exhibit charge and size selectivity depending on the structure of the interacting proteins. It is, therefore, important to understanding the molecular-level interactions that constitute the tight junction architecture, especially at the BBB interface to enable the design of drug delivery modalities to treat neurological diseases. Experimental investigation of membrane protein association is challenging due to the hydrophobic membrane environment surrounding the proteins. Although numerous computational methods have been put forth to overcome this challenge, a comprehensive method capturing details of all possible states of protein interactions is yet to be developed. In this work, we present the development of a—new tool—that provides quantitative estimates of membrane protein association for generating a comprehensive Protein Association Energy Landscape (PANEL). The PANEL is a computationally inexpensive approach to producing energy profiles for a wide range of proteins. We have used our method to show the association of membrane proteins responsible for the selectivity of the blood-brain tight junctions.

Copyright © American Institute of Chemical Engineers. All rights reserved.

575187 A Bayesian Approach to Model Selection and Parameterization for Non-Bonded Interactions

We explore the usage of Bayesian inference as a force field parameterization paradigm and the techniques required to make these Bayesian decisions effectively. Current gradient-based strategies allow for the optimization of certain parameters, but have difficulty effectively sampling multimodal or shallow distributions. Another critical deficiency of these methods is their inability to compare multiple functional forms. Deciding which functional forms or levels of complexity are required will be an important consideration in developing the next generation of biomolecular force fields.

Reversible jump Monte Carlo (RJMC) facilitates equilibrium sampling of multiple models and their respective parameters, allowing for the estimation of probability distributions for both models and parameters. Statistical evidence in the form of Bayes factors can be taken from these distributions and allows for comparison of the relative fitness of force fields. We test this strategy on previously developed physical property surrogate models for the 2-center Lennard Jones plus quadrupole (2CLJQ) force field, allowing for fast comparison between surrogate model outputs and experimental data. RJMC sampling shows that, while the variable bond length of the 2CLJQ force field is justified for the studied diatomics and other similar molecules, the quadrupole parameter is not justified for use in the reproduction of physical properties such as density, saturation pressure, and surface tension. This analysis also shows that for other molecules, with larger expected quadrupole moments, the quadrupole parameter is required to reproduce these physical properties. This study shows the ability of RJMC to distinguish between cases where a certain parameter or functional form is justified or not.

In order to generalize this strategy, we also propose and test techniques to generate surrogate models "on-the-fly" for more general sets of parameters, using a multi-fidelity approach of equilibrium simulation, mixture reweighting with MBAR, and Gaussian process modeling to generate smooth surfaces of physical properties as a function of force field parameters. These surfaces can then be used in Bayesian inference via Monte Carlo sampling. We discuss the rational design of a multi-fidelity sampling hierarchy, and demonstrate the utility of this technique via sampling van der Waals interaction parameters for small molecules by comparing surrogate model outputs to experiment for simple physical properties (such as density, dielectric permittivity, enthalpies of mixing). We also discuss challenges associated with scaleup and the techniques required to overcome them, as well as future plans for using these techniques to rationally design better small-molecule biomolecular force fields, as a part of the Open Force Field Initiative.

Copyright © American Institute of Chemical Engineers. All rights reserved.

575337 Modulating Protein Conformational Dynamics Via Salt-Bridging Interactions

The sequences and three-dimensional structural folds of homologous proteins often fail to explain many striking differences in their functional characteristics, which are attributed to their dynamics in the solution state. The differences in dynamics of closely-related proteins present significant hurdles for structure-based design of small molecules targeting such proteins. In this work, we have studied using atomistic molecular dynamics (MD) simulations the interplay of protein dynamics and their role in affecting binding affinities of ligands in a family of signaling proteins. We hypothesized that protein flexibility can be modulated by incorporating or removing salt-bridging interactions, thereby designing new protein constructs where inhibitor potency can be enhanced or reduced in a predictable way. Presented in this work will be results where we identified several such salt-bridging interactions across various isoforms of signaling proteins and studied mutant proteins that revealed a reduction in potency of inhibition on reducing the protein flexibility.

Copyright © American Institute of Chemical Engineers. All rights reserved.

575380 The Use of Experiment Directed Simulations to Improve the Accuracy of Simulations

Experiment Directed Simulation (EDS) is a single replica biasing technique that is used to improve the agreement between simulations and experiments. Experiment Directed Simulation is a maximum entropy biasing technique that makes the average of simulated collective variable agree with a reference which usually comes from experiments. Here, we present a broad overview of EDS applications in peptide modeling in GROMACS simulator and EDS implementation in a modified simulation engine, HOOMD-TF. EDS had been used to improve the simulations of water molecules, lithium ions, and coarse-grained proteins in the past. We used EDS to improve all-atom simulations of GYG peptide and Aβ21−30 peptide by matching chemical shifts. Aβ21−30 peptide is 10 residue long fragment of amyloid peptide (Aβ42). We also used EDS in coarse-grained modeling of Aβ42 amyloid peptide self-assembly. The amyloid peptides are interesting because they are linked to Alzheimer's Disease that which has many variations and does not have potent cure. By adding experimental results directly to simulation with EDS, we were able to better characterize and understand structure and dynamics of these peptides. We recently developed HOOMD-TF software package, that combines GPU-accelerated computational, statistical, and analytical capabilities of TensorFlow (TF) library with the simulation engine called HOOMD-blue. We implemented EDS in HOOMD-TF, for making positions of Lennard Jones particles to follow special geometry in HOOMD. We showed that EDS can be applied in two simulation engines to simulate properties and behaviors of molecules that agree with experiments.

Copyright © American Institute of Chemical Engineers. All rights reserved.

575420 Developing First-Principles Based Interatomic Potentials with Bayesian Inference

Fundamental understanding of the electronic and catalytic properties of metallic nanoparticles and surfaces has been widely studied using density functional theory (DFT) calculations over the last few decades. However, the high cost of computations has hindered employing first principles DFT calculations for a large number of atoms (over a thousand) in real-world catalyst systems. Many highly efficient methods of calculating the energetic and structural properties of atomistic systems have been developed to overcome the size limitation of the DFT method by employing interatomic potentials. One of the most commonly used potential formalism for metals is the embedded-atom method (EAM). In standard approaches, EAM potentials have been developed by fitting parameters of its functional to the data from experimental observations and/or theoretical calculations and single optimal values are determined for each parameter. The current fitting techniques do not represent the uncertainty about the accuracy of the developed interatomic potential which is the result of fluctuation in fitted parameter values. To attain an estimate of the errors, we took advantage of the Bayesian model calibration method. Bayesian learning of interatomic potentials from ab initio data while considering the impact of fluctuations of these parameters is promising as a general, robust approach for force field optimization. This can be used in studying the uncertainties of atomistic simulations of metallic systems such as the catalytic properties of Au nanoparticles, particularly at the grain boundaries that are important for electrocatalytic reactions. With the aid of EAM potentials developed here, we studied the structure-activity relationships of grain boundaries of gold NPs for CO_2 reduction. The accurate prediction of *CO and *COOH adsorption energies using the developed structure–reactivity relationships helped us in observation of reduced overpotential for CO_2 electroreduction reaction on grain boundaries region. This superior activity which can be attributed to the induced strain by dislocations was also captured experimentally.

Copyright © American Institute of Chemical Engineers. All rights reserved.

575747 Free Energy Calculation on Human ß Defensin Type 1 in Different Redox Conditions

Human ß Defensin type 1 (hBD-1) is a small cationic peptide with three intramolecular disulphide bonds. It is constitutively produced by various epithelial tissues as well as induced in keratinocytes of human skin. It exhibits antimicrobial activity against several microorganisms, but the activity is salt sensitive. Compared to other defensins, it has minor antimicrobial activity in its oxidized form (wildtype) but much stronger activity in reduced form (analogue). In this project, the membrane-translocation capability of hBD-1 in both wildtype and analog forms will be predicted by investigating their free energy barriers crossing three different types of cell membranes: human red blood cell, gram-positive and gram-negative bacteria cells. The human red blood cell membrane is represented by 1-palmitoyl-2-oleoyl-sn-glycero-3-phosphocholine lipid bilayer, while gram-positive and gram-negative bacterial membranes by 1-palmitoyl-2-oleoylglycero-3-phosphoglycerol and 1-Palmitoyl-2-oleoyl-sn-glycero-3-phosphoethanolamine lipids in 3:1 and 1:3 molar ratios respectively. Umbrella-sampling method along with coarse-grained technique will be applied to calculate the potential of mean force on hBD-1 crossing different types of lipid membranes. The result will help to uncover the antibacterial dependence of hBD-1 on the redox conditions.

Copyright © American Institute of Chemical Engineers. All rights reserved.

575881 Molecular Dynamics Simulation on Human Beta-Defensin Type 3 Binding with CXCR4 Receptor

The protein human beta-defensin type 3 (hBD-3) is a component of the human innate immune system, mainly secreted from human epithelial tissues and mucosa. It is known to have multiple roles, including bactericide, binding with chemokine receptors (notably CXCR4), and is involved in the inhibition of HIV-1 replication. Structural changes can modify the chemotactic activities of hBD-3. Many factors govern its structure, especially its six cysteine residues that are capable of forming three pairs of disulfide bonds. The molecular detail of hBD-3's diverse chemotactic function is not yet known. All-atom molecular dynamics simulations using NAMD software were performed on hBD-3 analogs of varying structures interacting with the CXCR4 receptor embedded inside lipid bilayers. The binding structure and dynamics of hBD-3 with the CXCR4 receptor will be studied, and the result may aid in understanding the relationship between hBD-3's structure and its chemotactic functions.

Copyright © American Institute of Chemical Engineers. All rights reserved.

575958 Atomistic Insight Toward Dynamic Interactions in Pegylated Bovine Serum Albumin

The emergence of biological therapeutics has created higher demand for novel materials which can improve the stability and activity of native proteins. With bio-compatible polyethylene glycol (PEG) polymers having been FDA approved for internal administration within humans, synthesized PEG based bio-conjugates are increasingly being used to enhance the efficacy of native protein drugs.

In this work, ten different initial orientations of the PEG polymer were generated – prepared by evenly distributing the PEG end-to-end vector over the accessible protein surface. This methodology was employed to adequately capture the conformational space, which increases in complexity as a function of polymer molecular weight, that the grafted PEG chain could sample around the protein. This methodology enabled us to collect microsecond scale atomistic molecular dynamics simulation data of PEGylated BSA using PEG molecular weights of 2, 5, 10 and20kDa. Simulations were conducted with the CHARMM c36m and CHARMM ether forcefields. In addition to exploring protein-polymer interactions, the effects of the chemical environment of the conjugated site were studied using N-terminal PEGylated BSA as well as Lys116 PEGylated BSA. Simulation data from N-terminal conjugated PEG showed a "shroud"-like wrapped conformation to be favored as the linear PEG chain molecular weight increases, which is in correlation with experimental data.[1] Calculation of PEG residence time near each residue in BSA revealed that lysine-PEG interactions play an important role in promoting PEG-BSA interactions (Figure 1). Due to the unique location of Lys116, results from Lys116 PEGylated BSA simulations demonstrated how the chemical environment of the conjugation site can affect dynamics of grafted PEG.

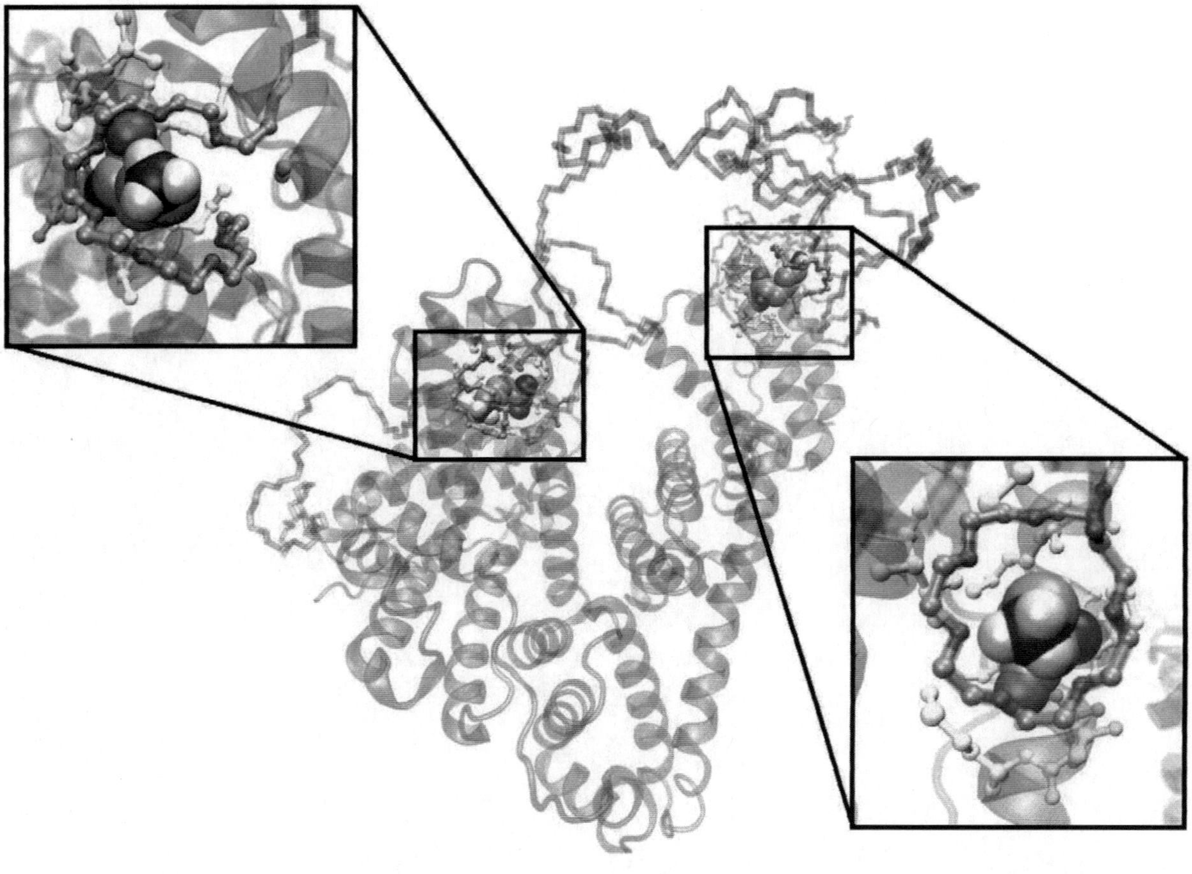

Figure 1: Representation of loop-like conformation formed by PEG near lysine residues in a 10 kDa PEGylated system when forming a "shroud" like conformation around BSA. Hydrogen atoms are not

shown for clarity. Surrounding residues near each lysine are colored in yellow (hydrophobic) and green (all others).

[1] Munasinghe, A.; Mathavan, A.; Mathavan, A.; Lin, P.; Colina, C. M. Molecular Insight into the Protein–Polymer Interactions in N-Terminal PEGylated Bovine Serum Albumin. *J. Phys. Chem. B* **2019**

Copyright © American Institute of Chemical Engineers. All rights reserved.

576170 Molecular Dynamics Modeling Based Investigation of Cryoconcentration of Solutes during Freezing and Its Impact on Protein Stability

Objective: Bulk protein therapeutics are generally frozen for long term storage to prevent product degradation, and facilitate product transportation. The freezing process of bulk protein solutions leads to cryoconcentration of solutes, which may impact protein stability. However, the underlying mechanism and molecular details of proteins unfolding and aggregating at the cryoconcentration microenvironment are poorly understood. Molecular dynamics (MD) simulations have been developed to investigate how cryoconcentration of different solutes affects protein stability during the bulk freezing process. The simulations results will be further validated with previous experimental data on solutes cryoconcentration acquired by our group.

Methods: The initial conditions applied into the MD simulations were selected based on previous cryoconcentration experiments in which 1 L bottles containing 600 mL of a protein solution of 0.5 mg/mL lactate dehydrogenase (LDH) in 10 mM phosphate buffer at pH 7.0 with or without 2.5% (w/v) sucrose were frozen at -30°C. Therefore, LDH, phosphate salts and sucrose molecules were added into the simulation box. The MD simulation trajectories and their analyses were carried out using GROMACS/VMD software. We applied both MARTINI and CHARMM force-field for all-atom as well as coarse-grained simulations, respectively. The initial coordinates for LDH were taken from the protein data bank entries (pdb: 5NQB). The protein was initially modeled using the psfgen tool in VMD, and then further solvated in a simulation box and charge-neutralized with $NaPO_4$. The box volume was optimized in the NPT ensemble by initially applying 500 cycles of a conjugate-gradient minimization scheme followed by a short 40-ps MD run in NPT assemble. The temperature was controlled using the Langevin thermostat, and the pressure was controlled by the Nose-Hoover barostat. The simulations were carried out using periodic boundary conditions. These briefly equilibrated systems of LDH protein were further subjected to MD simulation sets in the NVT ensemble. Computations were performed on HPC, a supercomputer cluster at the University of Connecticut.

Results and Discussion: The cryoconcentration effect was evaluated applying a model protein solvated in a simulation box of aqueous solvent in which different ions and stabilizers were added for case studies. Analyses of the resulted trajectories reveal the effect of temperature, ionic strength, and solutes concentration on the protein unfolding state, biomolecular interactions, and protein aggregation.

Copyright © American Institute of Chemical Engineers. All rights reserved.

576223 High-Pressure Phenomena in Adsorbed Films: A New Route to an Experimental Determination of Effective Tangential Pressure

High-pressure phenomena have been observed in many studies of adsorbed films on strongly wetting solid substrates. These include chemical reactions that normally require a high pressure (e.g. 10,000 bar or more), the formation of high-pressure solid phases and spectra indicating strong compression of the film. While it is possible to measure the pressure acting normal to the substrate surface, P_N, no method has been developed so far to measure the tangential pressure, P_T, parallel to the surface. However, this tangential pressure is of particular interest, since it is usually much larger than the normal pressure, and is the main driving force for high-pressure phenomena in these films.

While the normal pressure is uniquely defined for a planar surface, the tangential pressure at a point **r** is not uniquely defined at the nanoscale. In the commonly used 'virial-route' the average of the intermolecular forces acting in the direction parallel to the surface is obtained. The resulting tangential pressure then depends on the fraction of these forces that are assigned to a particular point in space, **r**. We show that by integrating $P_T(\mathbf{r})$ over a small region of space, roughly the range of the intermolecular forces, we obtain an effective tangential pressure that is unique and well-defined. The grand canonical Monte Carlo simulation results show that this characteristic integration range has negligible correlation with the wetting parameter, which denotes the strength of the solid-fluid interaction, and the characteristic length remains as a constant for large pores. Further investigation reveals that the characteristic length of convergence is related to the density distribution of adsorbed layers, which explains its weak dependence on the wetting parameter. Based on this characteristic length, we propose a new route, the '2D-route', to the effective tangential pressure free from the ambiguities in the definition of the 'virial-route' local tangential pressure. The only input parameter for this '2D-route' is the molecular cross-sectional area. In practice, we can obtain the input parameter by combining standard experimental adsorption data with a 2D equation of state derived from statistical mechanics. We expect this '2D-route' to be the first experimental route to the pressure enhancement in the adsorbed layers.

Copyright © American Institute of Chemical Engineers. All rights reserved.

576650 Applications of *Freud* for Nanoscale Simulation Analysis

Although particle simulations span a wide range of length and time scales, most existing standardized analysis tools are strongly focused on biomolecular simulations, and few tools exist for colloidal and coarse-grained simulations. This talk will showcase *freud*, a Python package that aids in calculating quantities that are frequently of interest in simulations on colloidal length scales. Perhaps the most prominent use-case we will highlight is the identification of phase transitions, including 2-dimensional melting, shape-driven solid-solid phase transitions, and depletion-driven crystallization. We will also demonstrate that *freud* has been used to study such diverse problems as the inverse design of isotropic pair potentials and the engineering of strain fields in colloidal crystals. A particular emphasis will be the ease of building new analyses using the core infrastructure of *freud*. The poster's final emphasis will be on how *freud* can be used in cutting-edge applications such as the computation of descriptors for machine learning algorithms or on-the-fly simulation analysis to study complex phenomena as they occur.

Copyright © American Institute of Chemical Engineers. All rights reserved.

576835 Building Computational Tools to Help Guide Experimentalists in the Discovery of New Solid Phases of Organic Materials

Due to the fact that crystalline structures of small organic molecules can vary in properties, fields such as as pharmaceutical formulation, organic electronics processing and explosives preservation have interest in fast and efficient methods to predict favorable structures. A failure to know all potential solid forms has lead to recalls, reformulation, or a missed opportunity to utilize a material to its fullest extent.

Common computational methods are great at predicting new structures, but fail to accurately rank the stability of predicted structures by neglecting temperature and entropic effects. Structure predictions have relied on lattice energy to determine the crystalline stability despite numerous experimental and computational studies that have shown that enantiotropic behavior is prevalent. In particular, one study has shown that 20% of a set of > 500 experimentally know polymorphs re-rank with temperature.

My studies have been focused on improving and testing computational techniques to efficiently determine the relative free energy stability of organic polymorphs. To date, I have looked at how the following factors affect polymorph stability for our methods: 1) the effects of thermal expansion, 2) the importance of anharmonic motions, 3) force field complexity, and 4) force field sensitivity. All tests have the end goal of making our methods faster and more accurate to eventually integrate the use of free energy as the criterion for stability in predicting new crystalline structures. Additionally, all techniques are available to use by those interested in two python based code packages that I manage.

Copyright © American Institute of Chemical Engineers. All rights reserved.

576889 Metal Organic Frameworks As Cargos for the Delivery of an Anti-Cancer Drug, Curcumin

Cancer is one of the leading causes of morbidity and mortality around the globe. Curcumin is an anti-cancer drug, which can be found in turmeric. It has been used in several preclinical and clinical studies with the goal of preventing and treating a great variety of different cancers. However, its very low solubility in water (estimated to be 3 ×10-8 μM) and chemical instability under alkaline conditions results in its poor bioavailability. Here we have used metal-organic frameworks (MOFs) as potential carrier to improve curcumin's solubility, chemical stability, bioavailability, and therapeutic efficacy. We have performed grand canonical Monte Carlo (GCMC) simulations of curcumin absorption in selected MOFs at 300 K and 1 atm. Charges on MOF and Curcumin were calculated by using Eqeq and DDEC methods, respectively. GCMC simulations were performed for 3 x 107 steps and were analyzed to estimate the drug adsorption capacity for different MOFs. Results from our study can be useful in developing next generation drug delivery carriers for curcumin and similar drug molecules.

Copyright © American Institute of Chemical Engineers. All rights reserved.

576930 Foyer: A Framework for Defining Force Field Usage Semantics and Atom-Typing Molecular Systems

The availability of force fields for molecular simulation has reduced the effort researchers must devote to the task of determining the interactions between species, allowing them to instead focus on the motivating scientific questions. However, determining which parameters in a forcefield to use is still often a very tedious and error prone task, given that force fields may contain hundreds of unique parameters for a single element, differentiated by the chemical context for which they apply. While tools for performing atom-typing exist (i.e., determining which parameters describe an atom's interaction potential with its environment based on its chemical context), they are often tied to a specific force field or simulation software and typically employ a rigid hierarchy where rules that identify more specialized atom types must be called ahead of more general atom types. More importantly, many of these atom-typing tools are not directly tied to the files containing the definitions of the force field parameters, which may introduce ambiguities as to appropriate usage and the source of parameters. The strong dependence of force field parameters on the chemical context of the atoms demands a codified, logic-based expression tied directly to those parameters.

Here, we present the Foyer tool [1,2,3] that eliminates ambiguity in force field parameter usage thereby enhancing reproducibility and simplifying the process of disseminating new force fields. Foyer provides a force field agnostic method for defining parameter usage and a Python based software library for evaluating these files. Foyer relies upon the OpenMM XML file format [4] to encode parameter usage, where this format is extended to allow the chemical context to be encoded using the SMARTS language for chemical patterns [5]. In Foyer, rule precedence is set by "overrides" statements encoded in the forcefield file, rather rigid rule hierarchies, allows rules to be embedded in any order within the actual force field files, thus delivering annotations that are both human and machine readable. Rules are evaluated using an iterative process, whereby all rules are fully evaluated, allowing the force field file to be validated for completeness. Foyer has been developed to integrate with the Molecular Simulation Design Framework (MoSDeF) [6].

1. Iacovella, C. R.; Sallai, J.; Klein, C.; Ma, T. Idea Paper : Development of a Software Framework for Formalizing Forcefield Atom-Typing for Molecular Simulation. In *4th Workshop on Sustainable Software for Science: Practice and Experiences (WSSSPE4)*; 2016.
2. Foyer [Online]. Available: https://github.com/mosdef-hub/foyer.
3. Klein, C., Summers, A.Z., Thompson, M.W., Gilmer, J., McCabe, C., Cummings, P.T., Sallai, J. and Iacovella, C.R., 2018. Formalizing Atom-typing and the Dissemination of Force Fields with Foyer. *arXiv preprint arXiv:1812.06779.*
4. OpenMM [Online]. Available: http://openmm.org
5. http://www.daylight.com/dayhtml/doc/theory/theory.smarts.html
6. https://github.com/mosdef-hub

Copyright © American Institute of Chemical Engineers. All rights reserved.

577066 Electrochemical Reduction of CO_2 on Transition Metal- P Block Compositions

Carbon dioxide is very stable under environmental conditions and reduction to hydrocarbons is an endothermic process, hence highly negative potential combined with good catalyst efficiency is required to reduce the activation energy. Electrochemical reduction of CO_2 to value-added hydrocarbon fuels has attracted many researchers in recent decades trying to develop non-precious catalysts with high energy efficiency which can lower the reducing potential without compromising with the performance of the catalyst as well as selectivity of the products.

The goal of computational catalysis is to be able to study the performance of a catalyst for a specific reaction under specific reaction conditions. Therefore, to control the activity of a certain catalyst, we should tailor the catalyst atom by atom. Our work mainly focuses on using computational modelling to design and study transition metal and metal oxides (non-precious catalysts) that can electrochemically convert CO_2 to light hydrocarbons which can satisfy all or at least some the characteristics of an effective catalyst as discussed above.

Current work include CO_2 electroreduction to CO, HCOOH, CH_3OH, CH_4 as products on MoO_3, WO_3 , ZnO (wurtzite and zincblende), and ZnS (wurtzite and zincblende) compositions. We calculate adsorption energies and build free energy diagrams from various reaction pathways to find the lowest energy pathway followed and different products that can be formed at a certain reducing potential on each of the composition. We build volcano plots for both CH_3OH and CH_4 to rank the catalyst activity.

Copyright © American Institute of Chemical Engineers. All rights reserved.

577086 Melting of Glucose Disaccharides Using Molecular Dynamic Methods

Equilibration studies of eight glucose disaccharides (a-maltose, ß-maltose, sucrose, trehalose, isotrehalose, neotrehalose, galactose, and lactose) were performed at solid state temperatures for each molecule and 1 atmosphere. Molecular dynamic simulations were performed at constant pressure using a modified GROMOS 45a4 united atom force field. The torsional parameters and Lennard-Jones parameters were modified for this study. To test the force field, the computed density and the glycosidic torsion angles F and Ψ and compared them to experimental results. In addition, the melting studies of a-maltose were performed at 1 atmosphere. The disaccharide melted at 1000 K way outside the melting range reported in experimental studies of 458 K to 460 K.

Copyright © American Institute of Chemical Engineers. All rights reserved.

577137 Systematic Coarse-Grained Mapping and Potential Learning

The choice of coarse-grained (CG) mapping operator and the derivation of the corresponding coarse-grained (CG) potentials are critical in determining how effective a CG model for a molecule is. Here we present a technique for extracting a CG mapping operator from a hierarchical graph which encodes multiple mapping operators. We demonstrated the used of relevant thermodynamic properties, like entropy, to automate CG mapping selection from the hierarchical graph. We have also added the capability of learning CG potentials on-the-fly via force-matching to a recently developed plugin, HOOMD-TF. The plugin integrates TensorFlow (TF), a Machine Learning platform, with the GPU-accelerated simulation engine, HOOMD-blue. We have compared the CG potentials learned by a neural network to those modeled using a basis-set. These systematic strategies of mapping operator selection and CG potential learning were implemented for methanol, alkanes and peptides.

Copyright © American Institute of Chemical Engineers. All rights reserved.

577180 Phase Thermodynamics and Ab Initio Simulations to Optimize Oxidation Resistance of High Entropy Alloys in Additive Manufacturing

Materials used in vehicle engines, space shuttles, etc. are often exposed to extreme oxidation conditions with temperature over 1000K. This application requires improved materials with increased oxidation resistance. One approach is to use additive manufacturing to build oxidation-resistant coatings on traditional alloy engine parts. Hence, there is a need to search for materials that have a combination of specialized mechanical properties (amenable to additive manufacturing) and thermal/chemical properties (high oxidation resistance and melting point). High entropy alloys (HEA), which contain four or more constituent elements, are one possible class of materials to address this challenge. For example, equimolar Mo-W-Al-Cr-Ti has high oxidation resistance [1], while equimolar Nb-Mo-Ta-W and V-Nb-Mo-Ta have high yield stress of 1058 MPa and 1246 MPa respectively [2]. However, the HEA search space is far too large for exhaustive experimental search methods to discover alloys with desired properties. In this work we show how thermodynamic methods like Calculated Phase Diagrams (CALPHAD) and ab initio calculations from Density Functional Theory (DFT) can correlate with oxidation resistance and additive manufacturing ability of high entropy alloys. Even these theoretical methods are too slow to directly search the entire composition space. We further demonstrate the use of machine learning to guide these calculations and accelerate the underlying *ab initio* methods.

References:

1. Gorr, B., Azim, M., Christ, H. J., Mueller, T., Schliephake, D., & Heilmaier, M. (2015). Phase equilibria, microstructure, and high temperature oxidation resistance of novel refractory high-entropy alloys. *Journal of Alloys and Compounds*, *624*, 270-278.
2. Senkov, O. N., Wilks, G. B., Scott, J. M., & Miracle, D. B. (2011). Mechanical properties of Nb25Mo25Ta25W25 and V20Nb20Mo20Ta20W20 refractory high entropy alloys. *Intermetallics*, *19*(5), 698-706.

Copyright © American Institute of Chemical Engineers. All rights reserved.

577315 Mosdef: A Python-Based Molecular Simulation and Design Framework

The Molecular Simulation and Design Framework (MoSDeF) [1] is an extensible set of Python tools focused on designing chemical systems, atom typing, and screening over large sets of such systems in a programmatic fashion. MoSDeF has been developed as an effort to improve the current state of reproducibility within the molecular simulation community. Currently, MoSDeF is built around two key pieces of software, *mBuild* [2,3] for designing systems of interest, and *foyer* [4,5] for applying the appropriate force fields to a molecular system. *mBuild* supports a wide range of systems due to the robust particle connection method and ability to declaratively express repetitious structures with small amounts of code. This leads to parameterization of many features of a system, for example, from terminal group chemistry of a polymer chain, basis particles within a crystal lattice, size and composition of a nanoparticle, and much more. *Foyer* is a tool for atom typing molecular systems that does not rely on rigid, hard-coded, rule hierarchies for atom type application. By separating the engine that applies the force field from the force field file itself, the atom typing process can be debugged much easier, force fields can more easily be extended, evolved, and disseminated with publications. MoSDeF has been designed to be extensible, through the development of a plugin architecture. Additionally, in combination with the workflow management framework Signac [6], MoSDeF allows for large-scale screening capabilities, while also maintaining reproducibility and human-readability.

References

1. MoSDeF [Online]. Available: https://github.com/mosdef-hub.
2. C. Klein, J. Sallai, T. J. Jones, C. R. Iacovella, C. McCabe, and P. T. Cummings, “A Hierarchical, Component Based Approach to Screening Properties of Soft Matter”, Foundations of Molecular Modeling and Simulation, **2016**, pp. 79-92.
3. mBuild [Online]. Available: https://github.com/mosdef-hub/mbuild.
4. Iacovella, C. R.; Sallai, J.; Klein, C.; Ma, T. “In Idea Paper: Development of a Software Framework for Formalizing Forcefield Atom-Typing for Molecular Simulation”, 4th Workshop on Sustainable Software for Science: Practice and Experiences (WSSSPE4), **2016**.
5. Foyer [Online]. Available: https://github.com/mosdef-hub/foyer.
6. Signac Framework [Online]. Available: https://signac.io

Copyright © American Institute of Chemical Engineers. All rights reserved.

577610 Modeling Inhibition of Biocatalysts By Ionic Liquids Using Metadynamics

Ionic liquid (ILs) are salts that are liquid at room temperature and possess a number of unique and tunable properties among which include; high dissolution of cellulosic biomass, and limited inhibition and degradation of certain enzymes when solvated in low concentrations.[1, 2] These properties are exciting in the field to biocatalysis, as they provide a stark contrast in stability and safety compared to the harsh organics currently implemented in industry. Similarly, the selection of unique cation/anion pairs may result in a number of tunable properties.[3] IL solvents provide a promising framework as we work towards developing a one-step process for the conversion of biomass into fuel (i.e., both the pretreatment of biomass and enzymatic hydrolysis occurring in a single batch process). However, little is known about the mechanisms by which ILs inhibit the function of many biocatalysts. With the goal of using simulations to guide the rational design of biocatalysts suspended in ILs, we have used molecular dynamics to understand their deactivation mechanisms. Of the many modes by which IL solvents may deactivate proteins, we are specifically interested in probing their effects on protein ligand binding.

To this end, we will present our work on a model system: a retaining cellulase (endocellulase 1) in complex with a cellotetraose molecule, solvated at varying concentrations of an aqueous IL (1-butyl-3-methylimidazolium chloride) solution. Preliminary results indicate that preferential solvation of cellulose in ionic liquids, and inhibition of important binding site residues by cations may result in a decreased affinity between cellulose cellulase binding. To overcome the time-scale limitations of classical simulations and further understand the energetic driving forces of these interactions, we have implemented the funnel metadynamics variant to increase sampling of ligand binding and unbinding events.[4] To this end, we hope to gain a better molecular understanding of how protein-solvent, ligand-solvent, and protein-ligand interactions are impacted in the presence of ILs.

References

[1] V. Jaeger, P. Burney, J. Pfaendtner, "Comparison of Three Ionic Liquid-Tolerant Cellulases by Molecular Dynamics", **Biophys. J.,** 2015, 108(4) 880–892.

[2] P. Engel, R. Mladenov, H. Wulfhorst, G. Jager, & A. C. Spiess. "Point by point analysis: how ionic liquid affects the enzymatic hydrolysis of native and modified cellulose", **Green Chem**., 2013, 12, 1959-1966.

[3] Wilkes, J. S. "Properties of ionic liquid solvents for catalysis", **J. Mol. Catal. A**, 2004, 11-17, 1381-1169.

[4]V. Limongelli, M. Bonomi, M. Parrinello, "Funnel Metadynamics as accurate binding free-energy method", **PNAS,** 2013, 110 (16) 6358-6363

Copyright © American Institute of Chemical Engineers. All rights reserved.

577727 Automated Implementation of Trappe Force Fields Using Mosdef

The Transferable Potentials for Phase Equilibria (TraPPE) family of force fields is known for balancing accuracy and computational efficiency and for its ability to predict thermophysical properties for a wide range of chemical compounds and state points. The TraPPE website (http://chem-siepmann.oit.umn.edu/siepmann/trappe) provides access to published functional forms, parameters, simulated properties, and other resources that assist researchers in the successful implementation and use of TraPPE force fields (including united-atom, explicit hydrogen, polarizable, and small molecule families). The Molecular Simulation Design Framework (MoSDeF) is a set of python-based tools enabling TRUE (transparent, reproducible, usable by others, and extensible) molecular simulations compatible with a variety of simulation softwares. Here we present an extension of the TraPPE website to generate Foyer XML files for united-atom models, as well as monomer and dimer structures provided in PDB format together with their intra- and intermolecular energies. These files can be used within MoSDeF to generate the necessary input files for a range of simulation engines (LAMMPS, Gromacs, HOOMD, Casandra, GOMC, and CP2K), and the structures and associated energies further provide validation of the force field implementation.

Copyright © American Institute of Chemical Engineers. All rights reserved.

577873 An Accurate Coarse-Grained Model of Poly(acrylic acid) with Explicit Solvent Models of DMF and Water

Understanding the role of solvents in altering the conformations of biomaterials, polymers, and other soft matters is of fundamental interest in materials science and engineering. Here, we have developed a coarse-grained (CG) model of poly(acrylic acid) (PAA) to study its configuration in the presence of explicit solvent models of water, dimethylformamide (DMF), and their binary mixtures. The CG PAA model was developed based on the propionic acid model, and then validated by comparing its density and glass transition temperature with corresponding experimental data. The CG PAA 30-mer (30-monomer) in pure water and pure DMF exhibited a globule-like and a coil-like state, respectively. This is consistent with results from AA molecular dynamics (MD) simulations. The averaged radius of gyration (Rg) of globule-like and coil-like CG PAA 30-mer chains was ~9 Å and ~13.8 Å, respectively. In binary solvent mixtures, the averaged Rg values increased gradually from ~9 to ~13 Å with the mass fraction of DMF increasing from 2.6 wt% to 80.2 wt%. Moreover, the radial distribution function of solvents around the polymer chain suggests that there is an enhanced ordering of DMF molecules at the polymer-solvent interface even at a low mass fraction of DMF. The coordination number of DMF molecules within the first solvation shell of PAA increases with increasing the DMF mass fraction. This indicates that DMF molecules form a shielding layer and protect PAA from water molecules.

Copyright © American Institute of Chemical Engineers. All rights reserved.

578167 Mechanistic Insights into Electrolyte Stability Toward Li/Na Anodes

Lithium (Li) and sodium (Na) metal batteries are considered as promising next-generation energy storage devices due to their ultrahigh energy densities. However, the high reactivity of alkali metal toward organic solvents and salts renders inevitable side reactions, which further leads to undesirable electrolyte depletion, cell failure, and evolution of flammable gas. A novel multiscale calculation, in combination with in situ optical microscopic observations, was applied to investigate the intrinsic instability of normal organic electrolytes (PC, DOL, DME, and TEGDME) towards Li and Na metal anodes. Once complexed with Li/Na ions, solvent molecules exhibit a significantly reduced the lowest unoccupied molecular orbital level, facilitating the electrolyte decomposition and gas evolution. Based on this mechanistic insight, novel strategies are proposed to build a stable Li/Na anode-electrolyte interphase and thus long-life safe Li/Na metal batteries.

Fig. 1. Comparisons among the LUMO levels of pure solvent and ion-solvent complexes.

Reference:

[1] Chen, X.; Shen, X.; Li, B.; Peng, H. J.; Cheng, X. B.; Li, B. Q.; Zhang, X. Q.; Huang, J. Q.; Zhang, Q. *Angew. Chem. Int. Ed.* **2018**, **57**: 734¨C737.

[2] Chen, X.; Hou, T. Z.; Li, B.; Yan, C.; Zhu, L.; Guan, C.; Cheng, X. B.; Peng, H. J.; Huang, J. Q.; Zhang, Q. *Energy Storage Mater.* **2017**, **8**: 194¨C201.

[3] Zhang, X. Q.; Chen, X.; Xu, R.; Cheng, X. B.; Peng, H. J.; Zhang, R.; Huang, J. Q.; Zhang Q. *Angew. Chem. Int. Ed.* **2017, 56**: 14207¨C14211.

[4] Zhang, X. Q.; Cheng, X.-B.; Chen, X.; Yan, C.; Zhang Q. *Adv. Funct. Mater.* **2017, 27**: 1605989.

[5] Zhang, R.; Chen, X. R.; Chen, X.; Cheng, X. B.; Zhang, X. Q.; Yan, C.; Zhang, Q. *Angew. Chem. Int. Ed.* **2017**, **56**: 7764¨C7768.

[6]. Cheng, X. B.; Yan, C.; Chen, X.; Guan, C.; Huang, J. Q.; Peng, H. J.; Zhang, R.; Yang, S. T.; Zhang, Q. Chem **2017**, **2:** 258¨C270.

Copyright © American Institute of Chemical Engineers. All rights reserved.

578176 The Transport Mechanism of CO_2/CH_4 Mixtures in Ionic Polyimides and Ionic Liquid Composite Membranes – MD Simulation Studies

The reduction and control of greenhouse gases emission is one of the most challenging problem that the society is struggling with nowadays. Greenhouse gases, like CO_2 and CH_4, are produced in many industrial processes and to limit their emission several pre- and post-combustion techniques have been applied [1]. However, these processes still face some limitations, like corrosion or volatility and toxicity of solvents used, i.e. monoethanolamine (MEA). Because of this, new materials or membranes are needed.

Ionic polyimides (i-PI) are a new class of polymers that are very promising as membranes to capture CO_2 [2,3]. This novel class of polymers contains ionic liquids species in the backbone separated by organic linker, which provides an additional degree of structural variety in the polymer system. Additionally to neat i-PI, significant attention is attracted by the composite membranes, which contain different ionic liquids (ILs) combined inside the neat polymer.

In our studies, the transport mechanism of two gases: CO_2 and CH_4 in neat and composite membranes were analyzed. For both gases, the diffusion in the neat i-PI is significantly slower that in the composite membranes with significant amount of ionic liquids added. This behavior results from plasticization effect of the ionic liquids on the rigid neat i-PI. However, addition of small, around 10 %, of IL to the system decreases the diffusion coefficient. This decrease in the gas diffusivity depends on the blocking effects of the gas diffusion route induced by the dispersed IL in the i-PI matrices rather than the plasticization effect.

References:

1. Martín, C. F.; Stöckel, E.; Clowes, R.; Adams, D. J.; Cooper, A. I.; Pis, J. J.; Rubiera, F.; Pevida, C. Hypercrosslinked organic polymer networks as potential adsorbents for pre-combustion CO_2*Journal of Materials Chemistry* **2011**, *21*, 5475.
2. Abedini, A.; Crabtree, E.; Bara, J. E.; Turner, C. H. Molecular Simulation of Ionic Polyimides and Composites with Ionic Liquids as Gas-Separation Membranes. *Langmuir***2017**, *33*, 11377-11389.
3. Abedini, A.; Crabtree, E.; Bara, J. E.; Turner, C. H. Molecular analysis of selective gas adsorption within composites of ionic polyimides and ionic liquids as gas separation membranes, *Phys.* **2019**, *516*, 71-83.

Copyright © American Institute of Chemical Engineers. All rights reserved.

578223 Understanding Temperature-Induced Self-Folding of Thermo-Sensitive Polymer Grafted Graphene

Physical and chemical properties of 2D materials like graphene are strongly affected by changes in their shape through the introduction of curvature, wrinkling, and folding. Graphene sheets can be folded to form 3D structures by grafting thermo-sensitive polymers to its surface and by changing the temperature. In this study, we attempt to unravel the mechanism of graphene self-folding by conducting coarse-grained (CG) molecular dynamics (MD) simulations of poly(N-isopropylacrylamide) (PNIPAM) grafted graphene models. PNIPAM is a classing thermo-sensitive polymer with a lower critical solution temperature (LCST) of 305 K. Below and above its LCST PNIPAM is in a coil-like and globule-like state. MD simulations were carried out below (290 K) and above (320 K) the LCST of PNIPAM in an aqueous medium. The CG MD simulations were performed at different grafting densities, coverages, and arrangements of the PNIPAM chains on the graphene surface. The simulation trajectories were analyzed to investigate the mechanism of graphene folding induced by coil-to-globule conformation changes of PNIPAM. The study further demonstrate possibilities of controlling the graphene folding to generate specific 3D structures.

Copyright © American Institute of Chemical Engineers. All rights reserved.

578245 Elucidating Biofuel Sooting Mechanisms: A Combined Reaxff and DFT Study

In this study, reactive molecular dynamics (RMD) is employed as an automated approach for proposing reaction
networks. Although the atomic forces for RMD would ideally be computed with high level
ab initio methods or even density functional theory (DFT), the nano- to millisecond time scales
that are required to observe combustion chemistry at moderate temperatures and pressures are
much too long for DFT-RMD to be feasible. By contrast, the reactive force field (ReaxFF)
renders classical RMD computationally tractable. ReaxFF simulations are performed for various biofuels to investigate

the soot formation mechanism. Subsequent to the RMD simulations, traditional DFT transition state calculations are used to validate the ReaxFF-predicted reaction pathways.

Copyright © American Institute of Chemical Engineers. All rights reserved.

576506 Molecular Dynamics Simulation on Interaction of Human Beta Defensin Type 3 with Different Lipid Bilayers

Human beta defensin type 3 (hBD-3) belongs to the human innate immune system. It is a natural antibacterial peptide mainly secreted from the human epithelial cells, and has a charge density of +11. It can interact with bacterial cell membranes which are overall negatively charged and can translocate through the cell membrane. In this project, the interaction and dynamics of hBD-3 in both monomer and dimer forms embedded inside different negatively charged lipid bilayers (to represent bacterial membranes) were investigated using CHARMM and NAMD molecular dynamics simulations. It was found that water molecules can cross the lipid bilayers through hBD-3, with the flow rate dependent on the lipid chemical composition and oligomerization form of hBD-3. Aggregation of negatively charged lipids around hBD-3 was observed based on micro-second long all-atom simulations, although such kind of lipid segregation was not observed in mixed lipid membrane systems only.

Copyright © American Institute of Chemical Engineers. All rights reserved.

Simulation Study of Macromolecules Propagation Mechanism in Heterogeneous Porous Media

Jianqiao Leng, Yang Zhao, Baojun Bai, Mingzhen Wei, Missouri University of S&T

Abstract

Polymer is one kind of macromolecules that commonly used in petroleum industry to improve oil recovery. Understanding the propagation mechanism of polymer in porous media is critical to a successful polymer-flooding project. However, the transportation of polymer through porous media is complicated especially when inaccessible pore volume (IPV) is considered. With literature review, many factors including fluid properties, rock properties and operations can influence behavior of IPV for polymer flooding. This paper collected the data from literature review and investigated the main factor that contributes to the IPV value analytically. The result shows that the solid entrapment may be the main reason conducting the IPV. Salinity, porosity and shear rate could be the most influencing factors on IPV value of specified polymer. Based on concept of IPV and the influencing factors, this paper also deployed a correlation model to the current simulator in oil industry.

Introduction

Polymer flooding has been proven as a cost-effective chemical-EOR method. It can improve sweep efficiency by controlling fluid mobility. However, researchers observed the existence of excluded connected volume when polymer flow through porous media even at very fine mobility ratio ($M<1$) (Dawson, 1972. Di Marzio, 1970). This fraction of connected pore volume is inaccessible for polymer molecules but accessible for solvent water, which is called inaccessible pore volume (IPV) for polymer. With reported, IPV values are very inconsistent and range from 0 to 48% of pore volume. As a result, optimization of IPV can additionally improve sweep efficiency of polymer flooding, which would result in a more favored polymer flooding efficiency.

The most agreed reason for IPV is summarized in (Dawson, 1972. Liauh, 1979. Shah, 1978), that polymer molecules sizes are larger when compared to solvent molecules and pore sizes. Furthermore, by adsorption on the rock and entrapment by pore throat during polymer flooding, polymer retention can further decrease the pore size. Another reason is that the depletion effect of large molecules builds up a depletion layer at the pore wall and makes the mass centers of polymer molecules cannot reach the pore wall (Chauveteau, 1981. Sorbie, 1989. Omari, 1989). Third reason is the unfavored entropic effects push polymers away from solid boundaries (Liauh, 1979. Di Marzio, 1965). However, the second and third reasons lose some credibility when polymer can adsorb on the rock surface.

Some researchers criticize these explanations for IPV. Because the polymer molecule diameter (HPAM in 3% NaCl brine) is typically 0.5~0.8 microns (Sorbie 1991) while average pore throat diameter for 470 md Berea sandstone was reported over 6.7 microns (Seright, 2006).

Clearly, the pore throat is more than enough for polymer molecule to pass through. However, researchers all agreed the existence of IPV, though mechanism for IPV is under debate. The author supposed that the hydrodynamic radius of polymer spherical coil should be counted as in-situ polymer radius instead of intrinsic radius. Because polymer is a kind of soft matter and its coils can elongate by several hundred times of their initial length under shear. Moreover, the solvent molecules can permeate into polymer colloids, swell the polymer solute and further increase the hydrodynamic radius polymer solution which is shown in figure 1 (M. Doi, 2013).

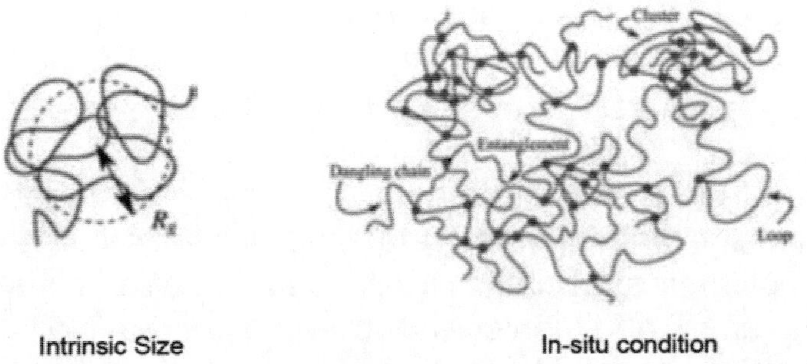

Figure 1. Real In-situ Condition Polymer Illustration vs Static Condition.

To prove this point, several researchers reported that the difference between polymer apparent in-situ viscosity and intrinsic viscosity measured by rheometer has strong relationship with IPV behavior (G. Chauveteau, 1984. Ferreira, 2018. Stavland, 2010). The result indicates that the hydrodynamic radius of polymer spherical coil could be much higher than static state molecule radius and cause the polymer unable to penetrate some part of connected pore volume.

Data Analysis

To analysis the sensitivity of each parameter, we collected IPV data from 132 experiments (Ref. 1-7, 17-26). The relationship between influencing factor and IPV value is plotted in cross plots shown in figure 2.

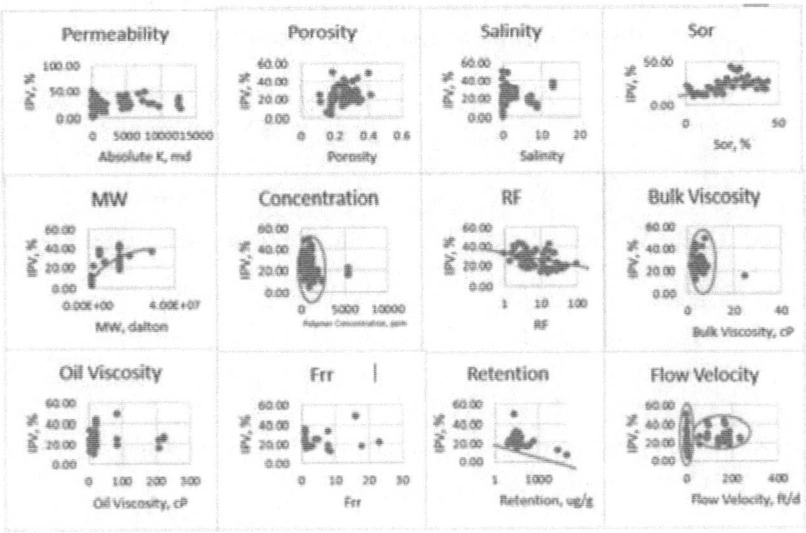

Figure 2. Crossplots for Infuencing Factors on Inaccessible Pore Volume

Based on the data set we collected from literature, we can obtain some analytical result listed below:

1. Sor can influence IPV value negatively.
2. Moleweight of polymer may have a positive effect on IPV value and the relationship is more likely semilog.
3. Resistance factor may have a negative effect on IPV.
4. The higher retention of polymer, the lower IPV could be.
5. Concentration and bulk viscosity of polymer tends to be very low for obvious IPV value.
6. Flow velocity (proportional to shear rate) falls at low velocity range and ultra high velocity range (>100 ft/d).

These observations tend to indicate that most of the experiments having a relatively low viscosity of polymer. Velocity range (shear rate range) also provide evidents for this point. Increasing shear rate, polymer especially HPAM can behave a shearthinning (decreasing viscosity at low shear rate range) and then shearthckening (increasing viscosity at medium shear rate range). The ultra high shear rate may most likely conduct a mechanical degradation for polymer, which can decrease polymer viscosity sharply. The velocity range falls at two separated parts at low velocity and ultra high velocity contributing to our inference.

The reason for this is simple to understand. Since IPV measurement needs polymer breakthrough earlier than solvent, if viscosity of polymer is much higher than solvent (eg. Water), the velocity advance results from IPV cannot overcome retarding by viscosity. The IPV may not be obvious to measure.

Correlation Model Based on Experimental Result

The concentration in a pore can be a function of pore size and polymer molecular weight. Several researchers investigated experiments for HPAM molecules distribution in APV zones based on ratio of pore radius to polymer molecule radius (Liauh, 1979. M Ali, 2017). Experimental data with same polymer size and different pore size core flooding is fitted using exponential function with 95% confidence bounds shown in figure 3.

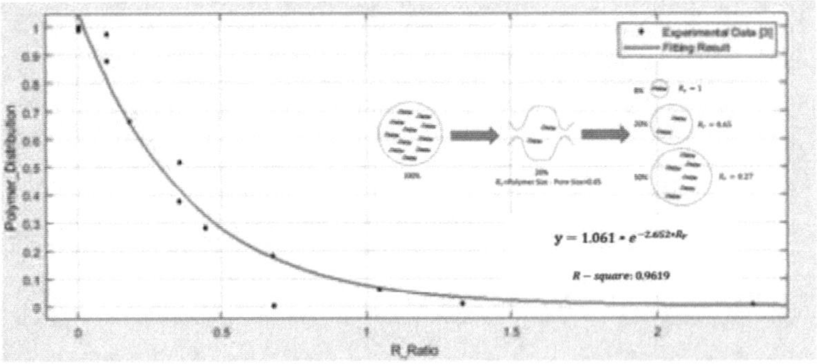

Figure 3. Polymer Molecules Distribution (PMD) vs Radius Ratio ($R_{polymer}$: R_{Pore})
(Marked as R_r)

The fitting result can be expressed in equation 1:

$$\text{PMD} = 1.061 * \exp(-2.652 * R_r) \ldots\ldots (1)$$

Based on the experimental results, more polymer molecules will flow into larger pores and less molecules flow into smaller pores. These experimental results can be illustrated using figure 6. Based on the same polymer molecule size, the larger the radius ratio Rr is, the less polymer can flow in the pores. For instance, 8% of polymer will flow through pore size with Rr=1, 20% of polymer will flow through pore size with Rr=0.65 and 50% polymer will flow through pore size with Rr=0.27.

As a result, with same flow rate, higher IPV will cause smaller APV and in result, a higher velocity is achieved by polymer component to keep material balance. At the same time, considering insensitive to IPV by water molecules, the concentration of polymer component will be lower. With this analysis, all three software cannot correctly reflect the IPV effect on polymer flow.

Based on experiments did by (Liauh, 1979), polymer molecules distribution result is shown in figure 6. Then polymer mole fraction flowing in-to the core can be expressed using equation 2:

$$X_L = \frac{\int_{R_{r,IPV}}^{R_{r,max}} PMD\, dR_r}{\int_{R_{r,min}}^{R_{r,max}} PMD\, dR_r} \ldots\ldots (2)$$

Where, X_L is mole fraction of polymer in the core. $R_{r,min}$ is minimum radius ratio. $R_{r,IPV}$ is critical radius ratio that polymer can propagate through the pore. $R_{r,max}$ is the maximum radius ratio.

Concentration in IPV is 0 and IPV is calculated assuming that pore radius (Rp) distributes evenly and polymer radius keep constant:

$$\text{IPV} = \frac{\int_{R_{r,IPV}}^{R_{r,max}} 1/R_r\, dR_r}{\int_{R_{r,min}}^{R_{r,max}} 1/R_r\, dR_r} \ldots\ldots (3)$$

IPV Sensitivity Analysis with Correlated Model

Mathematical Model

A correlation for concentration distribution is set up. The core model used is illustrated in figure 4.

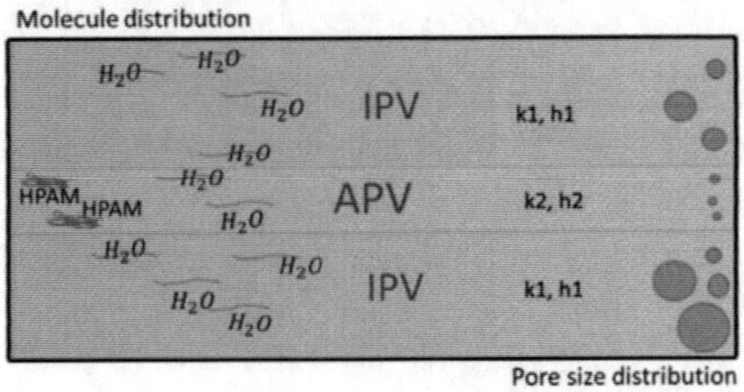

Figure 4. 1D Core Model with Consideration of IPV

Assuming: Only one layer in the core contains IPV, which means some of pore volume is not accessible to polymer, and APV where is freely accessible to polymer; Incompressible fluid flow; Darcy's law equation applied; No movable oil considered; No crossflow. A fractional flow model that is used to conduct sensitivity analysis is shown below.

$$\begin{cases} \nabla(v_w f_w) + \dfrac{\partial(\emptyset S_w)}{\partial t} = 0 \\ \nabla(v_p f_w C_p) + \dfrac{\partial(C_p \emptyset_p (S_p))}{\partial t} = 0 \\ \nabla(v_w f_w C_t) + \dfrac{\partial(\emptyset C_t S_w)}{\partial t} = 0 \end{cases} \quad \dots\dots (4)$$

B.C.: $\frac{\partial(v f_w C)}{\partial y} = \frac{\partial(v f_w C)}{\partial z} = 0 \; for\ all\ time\ C|_{x=0} = C_{inlet}\ C|_{t=0, x\neq 0} = 0$

Where, $fw = \frac{M_w}{M_w + M_o}, M\ is\ phase\ mobility.\ \emptyset_p = \emptyset * (1 - IPV), S_p = S_w * (1 - IPV)$, Cp is polymer concentration, Ct is tracer concentration.

Model Description

Table 1. Specification of the Correlated Model for Sensitivity Investigation

Layer Num	Model Length, cm	Cross-section Area, cm2	Porosity	Permeability, md	Sor
1	4.4	4.5	0.329	100	0.355
Polymer Concentration, ppm	Tracer Concentration, ppm	Polymer MW, Dalton	Layer Thickness, cm	Model Width, cm	Permeability, md
1800	1000	1.8E7	3	1.5	100

Assumptions

In this sensitivity analysis model, minimum radius ratio is assumed as 0.1, and maximum radius ratio is assumed as 2.5 in the sample core. Specification of the model is listed in table 1.

Result

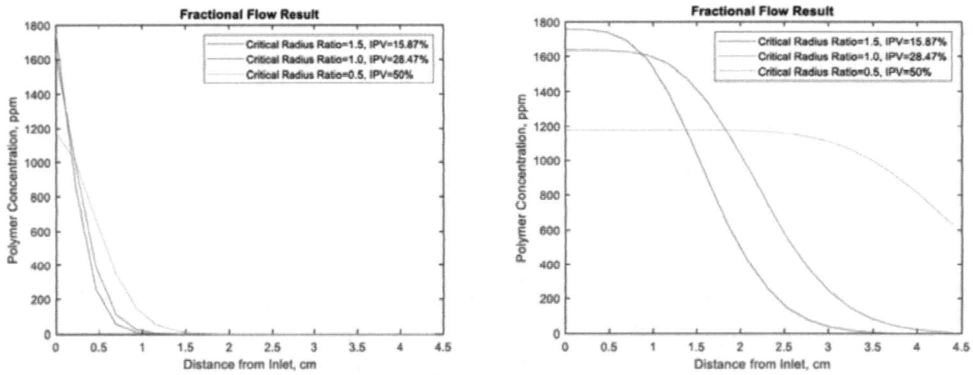

Figure 5. Concentration Profile after 1 min (left) and 10 min (right) of Polymer Flooding

Three cases with $R_{r,IPV} = 1.5,\ 1,\ 0.5$ respectively were investigated using the correlated model. The corresponding IPV for three cases are 15.87%, 28.47%, 50% which is calculated using equation 15. The results for polymer profile distribution along the core direction after 1 minute and 10 minutes of polymer flooding are shown in figure 5. Injection rate is assumed constant at 0.1 cc/min.

The result indicates that the polymer front moves faster for higher IPV value case than lower IPV value case. At the same time, the concentration for polymer is decreased with decreasing IPV value. It means that the model can treat IPV effectively on both polymer flow velocity and concentration sensitivity.

With experiments using sandpack polymer core flooding (eg. Ferreira, 2018), we noticed that for 1800 ppm HPAM, the adsorption could be as high as 1.9 $\mathrm{mg/g}$ and max Frr could be 4. This result is implemented in the correlated model. The result of Frr along with changing IPV are shown in figure 6.

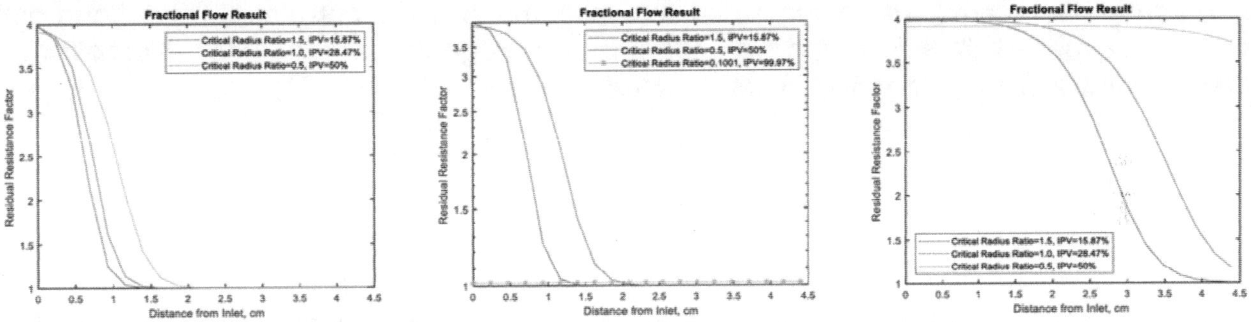

Figure 6. Residual Resistance Factor at 1 min (Left & Middle) and 10 min (Right) of Polymer Flooding with Varied IPV

Difference on Max Frr results among three cases are not apparent. This is because the adsorption curve for polymer concentration range from 1200 ppm to 1800 ppm varies very limited shown in figure 6. The influencing front velocity for polymer residual resistance factor is very sensitive to IPV. Frr can varied a lot when IPV is very high. This can be shown compared with extreme case in figure 6 (right), where three cases with $R_r = 1.5, 0.5\ and\ 0.1001$ respectively are investigated. The IPV for three cases are 15.87%, 50% and 99.97%.

The case with IPV=99.97% refers to matrix where polymer can be almost denied from flow in. The star line in figure 11 shows this case Frr distribution, which shows nearly no Frr can be generated in this kind of formation. As a result, for formation with relatively large IPV, very limited damage can be done to this formation by polymer flooding. This result can be very beneficial for conformance control using polymer or gel treatment and the result can match the experimental data better than commercial software result.

Conclusion

Inaccessible pore volume has been a key influencing factor for successful conformance control in petroleum industry. With reported, IPV values are very inconsistent and range from 0 to 48% of pore volume. Optimization of IPV can improve sweep efficiency of polymer flooding, which would result in a more favored polymer flooding efficiency. But ignoring the effect of IPV on concentration and retention can result in inaccurate prediction on conformance control efficiency.

Data analysis indicates a relatively positive attributes of Sor, MW, Porosity to IPV, and negative attributes of RF, Frr to IPV. Low viscosity range of IPV indicates the measurements need a close viscosity of polymer to water solvent.

Results of sensitivity analysis indicate that the higher the IPV is, the higher velocity of polymer front and lower concentration of polymer contained in the core. Meanwhile, Frr is sensitive to IPV, and for low permeability zone with ultra-high IPV value, the polymer flooding can barely damage the low permeability zone because Frr in low permeability zone is nearly ignorable.

The correlated model built in this paper can fully describe the polymer concentration change related to inaccessible pore volume and velocity difference based on molecule size to pore size ratio. Compared with commercial software, this paper's correlated model can better represent polymer distribution and retention in heterogeneous reservoir. Moreover, the correlation can be easily integrated in commercial software.

Reference

1. Di Marsio, E. A. and C. M. Guttman, "Separation by Flow", Macromolecules, 3 (2), 131-146 (1970).
2. Shah, B. N., Lawrence, G., Willhite, P., & Green, D. W. (1978, January 1). The Effect Of Inaccessible Pore Volume On The Flow Of Polymer And Solvent Through Porous Media. Society of Petroleum Engineers. doi:10.2118/7586-MS
3. Chauveteau, G., Zaitoun, A. "Basic Rheological Behavior of Xanthan Polysaccharide Solutions in Porous Media: Effects of Pore Size and Polymer Concentration". Proceedings to the European symposium on EOR, (1981), pp 197-212.
4. Sorbie, K. S. (1989, January 1). Network Modeling of Xanthan Rheology in Porous Media in the Presence of Depleted Layer Effects. Society of Petroleum Engineers. doi:10.2118/19651-MS
5. Omari, A., Moan, M., Chauveteau, G. "Wall effects in the flow of flexible polymer solutions through small pores" G. Rheol Acta (1989) 28: 520. https://doi.org/10.1007/BF01332923
6. Di Marzio, E.A. 1965. Proper Accounting of Conformations of a Polymer near a Surface. *J. Phys. Chem.* 42: 2101-2106.
7. Trushenski, S. P. and S. P. Oupta, "Micellar Flooding—The Design of the Polymer Mobility Buffer Bank", Society of Petroleum Engineers Journal, 18(1), 5-12 (Feb., 1978).
8. Kolodziej, E. J. (1988, January 1). Transport Mechanisms of Xanthan Biopolymer Solutions in Porous Media. Society of Petroleum Engineers. doi:10.2118/18090-MS
9. G.Chauveteau, M.Tirrell, A.Omari, "Concentration dependence of the effective viscosity of polymer solutions in small pores with repulsive or attractive walls" Journal of Colloid and Interface Science Volume 100, Issue 1, July 1984, Pages 41-54, https://doi.org/10.1016/0021-9797(84)90410-7

10. Ferreira, V. H. S.1* and Moreno, R. B. Z. L.2*, “Rheology-based method for calculating polymer inaccessible pore volume in core flooding experiments” International Symposium of the Society of Core Analysts, SCA2018-026, August 2018.
11. Stavland, A., Jonsbraten, H., Lohne, A., Moen, A., Giske, N. H. "Polymer Flooding – Flow Properties in Porous Media versus Rheological Parameters", SPE EUROPEC/EAGE Annual Conference and Exhibition, Barcelona, Spain, (2010), pp 14–17.
12. Southwick, J.G., Jamieson, A.M., and Blackwell, J.: "Conformation of Xanthan Dissolved in Aqueous Urea and Sodium Chloride Solutions, "Carbohydrate Research (1982) 99,117-27.
13. Tang, Engao. Zhang, Songxian. Yang, Junru. Sun, Fujie. “Inaccessible Pore Volume Influence on Polymer Solution Fluid Flow in Porous Media.” Petroleum Geology and Recovery Efficiency. July. 2009, Vol. 16, No. 4. 1009-9603(2009)04-0080-03. https://wenku.baidu.com/view/e296edf77c1cfad6195fa785.html
14. M Ali and H Ben Mahmud. “The effects of Concentration and Salinity on Polymer Adsorption Isotherm at Sandstone Rock Surface” IOP Conference Series: Materials Science and Engineering 78 (2015) 012038 doi:10.1088/1757-899X/78/1/012038
15. M. Doi. Soft matter physics. Oxford University Press, 2013.
16. CMG STARS User Guide, version 2016.
17. Dawson, R., & Lantz, R. B. (1972, October 1). Inaccessible Pore Volume in Polymer Flooding. Society of Petroleum Engineers. doi:10.2118/3522-PA
18. Liauh, W. C., Duda, J. L., & Klaus, E. E. (1979, January 1). An Investigation of The Inaccessible Pore Volume Phenomena. Society of Petroleum Engineers. doi:NA
19. Pancharoen, M., Thiele, M. R., & Kovscek, A. R. (2010, January 1). Inaccessible Pore Volume of Associative Polymer Floods. Society of Petroleum Engineers. doi:10.2118/129910-MS
20. Knight, B. L., & Rhudy, J. S. (1977, October 1). Recovery of High-Viscosity Crudes By Polymer Flooding. Petroleum Society of Canada. doi:10.2118/77-04-07
21. Vela, S., Peaceman, D. W., & Sandvik, E. I. (1976, April 1). Evaluation of Polymer Flooding in a Layered Reservoir with Crossflow, Retention, and Degradation. Society of Petroleum Engineers. doi:10.2118/5102-PA
22. Dominguez, J. G., & Willhite, G. P. (1977, April 1). Retention and Flow Characteristics of Polymer Solutions in Porous Media. Society of Petroleum Engineers. doi:10.2118/5835-PA
23. Dabbous, M. K. (1977, October 1). Displacement of Polymers in Waterflooded Porous Media and Its Effects on a Subsequent Micellar Flood. Society of Petroleum Engineers. doi:10.2118/6203-PA
24. Osterloh, W. T., & Law, E. J. (1998, January 1). Polymer Transport and Rheological Properties for Polymer Flooding in the North Sea. Society of Petroleum Engineers. doi:10.2118/39694-MS
25. Hughes, D. S., Teeuw, D., Cottrell, C. W., & Tollas, J. M. (1990, February 1). Appraisal of the Use of Polymer Injection to Suppress Aquifer Influx and to Improve Volumetric Sweep in a Viscous Oil Reservoir. Society of Petroleum Engineers. doi:10.2118/17400-PA
26. Gupta, S. P. (1978, February 1). Micellar Flooding the Propagation of the Polymer Mobility Buffer Bank. Society of Petroleum Engineers. doi:10.2118/6204-PA

578134 Protein Corona on a Gold Nanoparticle Studied with Discontinuous Molecular Dynamics and Atomistic Molecular Dynamics Simulations

Multiple proteins adsorbed on nanoparticles (NPs) give rise to protein corona and alter NPs' surface properties. Gold nanoparticles have been widely used in biomedicals. Fundamental studies of protein corona formation on gold nanoparticles in an aqueous environment at the molecular and atomistic levels have critical applications in biodetection using surface-enhanced Raman spectrum as well as drug delivery. To achieve computational efficiency, highly efficient event-driven discontinuous molecular dynamics (DMD) simulations with atomistic resolution were first performed in an implicit water environment to simulate multiple protein adsorption on a gold NP. Subsequently, Atomistic molecular dynamics (MD) simulations in explicit water were conducted to study the protein-NP's interactions in the higher resolution. The combination of DMD and MD simulations at different scales estimated the packing structure and orientations of the adsorbed proteins, packing density and adsorption isotherm on a gold NP, as well as the proteins' secondary structure alternation.

Copyright © American Institute of Chemical Engineers. All rights reserved.

578709 Wetting Behavior of Imidazolium Based Ionic Liquids: A Molecular Dynamics Study

Aqueous ionic liquids due to unique properties [1,2] are of special interest, in-depth understanding of the structure–property relation of the systems is therefore desirable. Molecular dynamics (MD) simulations are performed to investigate the wetting behavior of aqueous imidazolium based ionic liquids (ILs), consist of the 1-ethyl-3-methylimidazolium ([EMIM]) cation combined with hydrophilic boron tetrafluoride ([BF4]) and hydrophobic bis (trifluoromethylsulfonyl)imide ([NTf2]) anion. To understand the influence of anion on solid-liquid interface of graphite surface together with varying the IL concentrations between 0 to 30 wt % are studied. The wetting behavior of both the IL droplets is characterized through a density profile, and orientation order paramete. The contact angle of aqueous IL decreases with increasing the ion-pair of IL in water. Due to the preferred adsorption of ILs, water starts to deplete from the surface with an increase in the ionic liquid concentration. Hydrophobic nature of ([NTf2]) anion drags the liquid towards the interface from the bulk. The nature of water-anion hydrogen bond networks, nature of anions, interaction with the surface are the determining factors in the wetting behavior of the aqueous ionic liquids

References:

(1) Bhargava, B. L.; Klein, M. L. Aqueous Solutions of Imidazolium Ionic Liquids: Molecular Dynamics Studies. *Soft Matter* **2009**, *5* (18), 3475–3480.

(2) Oprzeska-Zingrebe, E. A.; Smiatek, J. Aqueous Ionic Liquids in Comparison with Standard Co-Solutes : Differences and Common Principles in Their Interaction with Protein and DNA Structures. *Biophysical reviews* **2018**, *10* (3), 809–824.

Copyright © American Institute of Chemical Engineers. All rights reserved.

578481 First Principles-Based Multiscale Atomistic Methods for Next-Generation Electrocatalysts

Recent dramatic developments in first-principles quantum mechanics (QM) methods have enabled accurate predictions for the properties of various systems. However, such QM methods are limited to several hundreds of atoms for dozens of picoseconds (ps) (10^{-12} seconds). This capacity may grow by an order of magnitude over the next few years, but pure QM methods will remain far from meeting the requirement to inform the continuum degrees of freedom (e.g. solvent effect in electrochemical systems), which require millions of atoms (10^6) and microseconds (10^{-3} seconds) of simulations. Thus, there is an enormous gap between the scale of current QM methods and that of the real applications. In my research, I **develop and validate computational tools** to fill this gap while simultaneously applying them to **developing new generations of materials**.

A great step toward achieving this goal is the development of novel hybrid QM/molecular mechanics (QM/MM) frameworks. However, applications of the most current QM/MM approaches have been limited by lacking flexible, reactive, and polarizable force fields (FFs) with variable charge transfer (particularly at the QM/MM boundary) and accurate coupling of QM and MM with minimal computational cost.

In this talk, I present a new generation of polarizable reactive FFs (**RexPoN**) that I have developed based entirely on QM with no empirical information. The resulting RexPoN FF has already made significant breakthroughs in discovering the new properties of materials that have been hidden to QM and experiment. Indeed, the simulations of water system using RexPoN FF lead **to the most accurate properties ever predicted by a FF** (and more accurate than DFT). At 298 K, RexPoN finds an average of 2.1 Strong Hydrogen Bonds (SHB) with an average lifetime of 93 fs. Connecting these SHBs leads to a one-dimensional polymer with occasional branches to sidechains. This established the revolutionary new paradigm for the structure of water as a **dynamic polydisperse branched polymer (DynPol).** In addition, there has been confusion and debates about the nature of anomalies in supercooled water near 230 K over the last 40 years. Using RexPoN FF, I show the changes in the structure and properties of liquid water as it is supercooled from 298 K to 200 K. Vibrational frequency and SHB calculations show that there is a **topological transformation in which water changes from one-dimensional DynPol above 230 K to a different structural network at 230 K.** We find that this topological transition accounts for the anomalous properties long associated with supercooled water.

In addition, I present a novel hybrid QM/MM framework called **RexPoN embedded QM** (**REQM**) with dual boundary regions. REQM dynamically communicates forces and energies between the RexPoN FF and QM regions via a python program. This allows accurate long-range interactions and polarization effects of the FF (for the application size solvent) to be included with the QM region to study complex reactions. The results of REQM simulations for the electrochemical CO_2 reduction on the Pt and Cu surface using explicit water solvent are validated against explicit DFT-D3 calculations. This means that REQM allows physical size and time scales to be used for charge flow and polarization, while crucial catalytic reactions and rate determining steps be determined by short but accurate QM (PBE flavor of DFT with D3 Grimme dispersion forces). The results of this research to reveal the real catalytic mechanisms in electrochemical processes should help the design of novel catalysts with much high performance than the traditional ones.

Copyright © American Institute of Chemical Engineers. All rights reserved.

568127 Development of a Generic Mathematical Optimization-Based Nanomaterial Design Toolkit

Advances in nanotechnology are enabling the synthesis of increasingly complex nanostructured materials [1]. Concurrently, there are significant efforts in the areas of density functional theory (DFT) and machine learning to predict which nanomaterials are expected to have superior properties [2]. However, these approaches constitute heuristic searches and provide no guarantees on the quality of identified designs. To address the lack of guarantees on global optimality, we have previously demonstrated several approaches for modeling nanostructured crystalline material design problems via mixed integer linear programming (MILP) [3,4]. Using our experiences from these past works, we attempt to abstract the search for the best nanostructured crystalline material, formalize this as a rigorous design problem, and provide a common optimization framework for choosing the optimal nanostructuring decisions.

We present a new Python-based software tool for generically formulating and solving nanostructured material optimization problems. The tool takes input information about the crystal structure of the material, general forms of simplified structure-function relationships, and an objective function from which it automatically casts an MILP model. The latter can then be solved directly via standard optimization solvers available through Pyomo [5], or via customized decomposition approaches.

Part of the IDAES Multi-Scale Process Systems Engineering Framework [6-7], the tool provides interfaces to input material information via several levels of detail and customization. In the simplest case, standard crystal data files and predefined structure-function relationships can be used to instantiate the models. Conversely, in the spirit of open-source code, we also provide access to the modelling elements from which users can write highly customized scripts for building materials optimization models. In this way, we aim to support a wide range of applications and user expectations. This tool aims to effectively lower the barrier for applying mathematical optimization by materials experts while simultaneously codifying the material design problems for further development by mathematical optimization researchers.

References

[1] Xia, Y., Y. Xiong, B. Lim, & S.E. Skrabalak. (2009). Shape-Controlled Synthesis of Metal Nanocrystals: Simple Chemistry Meets Complex Physics? *Angewandte Chemie International Edition, 48*(1), 60–103.

[2] Saal, J. E., S. Kirklin, M. Aykol, B. Meredig, & C. Wolverton. (2013). Materials design and discovery with high-throughput density functional theory: The open quantum materials database (OQMD). *Jom, 65*(11), 1501–1509.

[3] Hanselman, C. L., & C.E. Gounaris. (2016) A mathematical optimization framework for the design of nanopatterned surfaces. *AIChE Journal, 62*(9), 3250-3263.

[4] Hanselman, C. L., D.N. Tafen, D.R. Alfonso, J.W. Lekse, C. Matranga, D.C. Miller, & C.E. Gounaris, (2018). Design of doped perovskite oxygen carriers using mathematical optimization. In Proceedings of the 13th International Symposium on Process Systems Engineering. San Diego, USA, 2461-2466.

[5] Hart, W.E., J. Watson, & D.L. Woodruff, "Pyomo: modeling and solving mathematical programs in Python." *Mathematical Programming Computation 3*(3) (2011): 219-260.

[6] Miller, D.C., J. Siirola, D. Agarwal, A.P. Burgard, A. Lee, J.C. Eslick, B. Nicholson, C. Laird, L.T. Biegler, D. Bhattacharyya, N.V. Sahinidis, I.E. Grossmann, C.E. Gounaris, and D. Gunter, "Next Generation Multi-Scale Process Systems Engineering Framework", Proceedings of the 13th International Symposium on Process Systems Engineering (PSE 2018), *Computer-Aided Chemical Engineering*, 44, pp. 2209-2214, Elsevier, Amsterdam, M. R. Eden, M. Ierapetritou and G. P. Towler (eds.) (2018).

[7] Institute for the Design of Advanced Energy Systems (IDAES) https://idaes.org/

Copyright © American Institute of Chemical Engineers. All rights reserved.

572254 Modelling Micellization of Rhamnolipid Biosurfactant By Mesoscale Simulation

The sustainability movement has been driving a desire to replace petroleum surfactants by biodegradable ones in many chemical industries. Designing eco-friendly products requires fundamental understandings of surfactant micellization at the molecular level. The present work aims to predict the self-assembling and micellar properties of rhamnolipid biosurfactants using mesoscale dissipative particle dynamics (DPD) simulations. A coarse-graining technique is developed here to address the issue of modeling molecules with complex structure by equal-sized DPD beads. The Rhamnolipid is dissected into coarse-grained particles of various diameters determined by the Connolly volume of their composing functional groups. "Linkers" are introduced to maintain the molecular structure by mimicking the conformations at the atomistic level. Nonbonded interaction parameters are sampled by particle insertion method and mapped to the activity coefficient of bead components. The devised framework preserves the efficiency and the simplicity of common DPD models, and improves the accuracy in describing chemical specifics resulting in more precise predictions for the micellization phenomena. The method is applied to systematically study rhamnolipids with different numbers of head and tail, as well as varied length of the tails (Rha-Cx-Cy, x & y = 8~16). Results obtained by simulations are envisioned to provide useful guidelines for future product design in broad industrial applications.

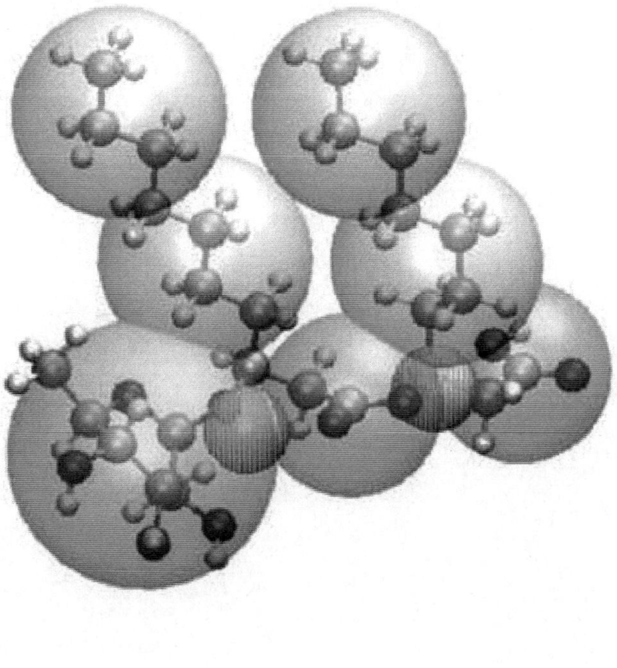

Copyright © American Institute of Chemical Engineers. All rights reserved.

578005 Oxidation State of Solutes in Molten Salts from Ab Initio Molecular Dynamics Simulations

Many electrochemical processes and applications use knowledge of an ion's oxidation state. For example, for very high-temperature molten salts, experimental determination of these oxidation states is costly, therefore an ab initio molecular dynamics simulation would be preferable. However, oxidation state is not a physical observable, has no quantum mechanical operator, and cannot be determined directly from AIMD simulations. We discuss a new method to identify the oxidation state of molten salt constituents using radial distribution functions of ab initio molecular dynamics simulations. While the oxidation states of akalis and halides in ionic liquids are typically intuitive, larger ions such as transition metals are much less so. This method is applied to ions using s,p,d, and f orbitals, and the accuracy of the results are discussed, and compared with existing methods and approximations such as an extensive analysis of molecular orbital and the Bader charge distribution analysis. The flexibility of this method for liquid phases was a major motivation for its development. Potential limitations of this method, such as solvent composition or solute concentration, are also discussed.

Copyright © American Institute of Chemical Engineers. All rights reserved.

572936 Quantum-Based Predictions for Network-Altering Chemical Reactions in Siloxane Polymer Systems

Chemical reactions involving the siloxane backbone can induce significant network rearrangements and ultimately degrade macro-scale mechanical properties of silicone components. High throughput atomistic modeling tools, such as semi-empirical quantum molecular dynamics, can yield detailed chemical understanding of these processes that is difficult to elucidate from experiments alone. Using large ensembles of simulations, we investigate the interconnections between mechanical loading, irradiation, and environmental moisture on chemical reactions in systems containing polydimethylsiloxane (PDMS) and polydiphenylsiloxane (PDPS). An approach to probe the chemistry underlying network rearrangements is proposed that couples graph-based structure recognition tools with efficient quantum simulations. Radiation-induced excitations of siloxane side groups are found to induce subsequent reactions resulting in intrachain cyclization and chain scissioning. A possible synergistic effect between water and radiation is identified that could promote alterations of a larger polymer network. Physical origins for increased chemical susceptibility in strained PDMS chains are proposed and explored in the context of hydrolysis reactions. Chain scission probabilities for PDMS under attack by water are extracted as functions of the backbone degrees of freedom, revealing complicated configurational inter-dependencies that increases the likelihood for hydrolytic degradation. This work was performed under the auspices of the U.S. Department of Energy by Lawrence Livermore National Laboratory under Contract DE-AC52-07NA27344. Approved for unlimited release, LLNL-ABS-771404.

Copyright © American Institute of Chemical Engineers. All rights reserved.

576483 Computational Corrections for Long Range Interactions in Periodic Simulations

While molecular simulations continue to provide useful insight into the behavior of chemical compounds and materials, they often neglect important energy and pressure components. Most simulations are carried out in a periodic cell as a computational expedient to avoid boundary effects. However, this approach requires a partitioning of the short-range and long-range energy and pressure components. Typically, binary interaction potentials are truncated to avoid double counting the pair images of interacting atoms. Then, the long-range energy component due to the interactions beyond the truncated potentials is calculated separately. The basic assumption behind this long-range correction is that the atomic pair distribution function has a value of unity beyond the truncation limit. With the exception of supercritical and near critical fluids, this assumption is reasonable as these long-range corrections typically constitute the significant portion of both energy and pressure components for these simulations. Given the complexity of this long-range correction, it is often neglected. Neglecting these corrections is reasonable in drug design because the bulk properties of the system are less important than the local ligand-receptor interactions. However, the bulk system is important in the prediction of fluid equilibrium and the design of materials typically pursued by chemical engineers. Some commercial force fields compensate for these long-range corrections, but the correction is dependent on the size of the simulation. These long-range interactions are analytically calculated for dispersion interactions, but the integrals for these corrections with electrostatic charges are not convergent and require reformulation of the charge distribution with a multipole expansion. Alternatively, inverse space summations may be employed such as Ewald summations, but these are conditionally convergent. To simplify the inclusion of these important long-range corrections, we have developed a simple empirical approach to estimate these energy and pressure corrections via the extrapolation of results from simulations of various sizes. We discuss the range of simulations for which this approach is accurate and provide examples to illustrate how it improves the results of bulk simulations.

Copyright © American Institute of Chemical Engineers. All rights reserved.

564643 Mixed-Metal Influence on Purity of Process Performance for Metal-Organic Frameworks with Open Metal Sites for Olefin/Paraffin Separation

Metal-organic frameworks (MOFs) with open metal sites (OMS) are known to be selective for ethylene relative to ethane. Our previous work demonstrated that the species of metal played a more important role than the topology and linker in determining the binding energies of small molecules on these open dimer metal sites. It suggested mixed-metal strategy is promising to tune the binding affinity of guest molecule in bimetallic systems. Additionally, the combination of molecular simulation and process modeling is very attractive and powerful to identify the performance of adsorbents in ethylene/ethane separation. First, we developed a Langmuir model to predict the isotherms of both mixed-metal MOFs and pure MOFs with OMS, then used them to estimate their process performance. The widely-used force field based-GCMC simulations is inaccurate to predict the adsorption isotherm in OMS MOFs because it doesn't account for the more localized interactions that can exist with OMS. In this work, we developed a dual-site Langmuir adsorption model to predict isotherms in OMS MOFs by dividing the total adsorption into chemisorption at OMS and physisorption at other sites. All model parameters were obtained by the combination of accurate density functional theory calculation and GCMC simulation. The prediction of a material process performance in practice is appealing. We developed an idealized PSA process to estimate their separation purity and further combined the isotherm model with the PSA process to identify whether mixed-metal MOFs with OMS is better than the pure MOFs with OMS. After a careful search, three pairs of mixed-metal MOFs have been demonstrated to improve the process purity. Additionally, the influence of the linker functionalization on the purity of mixed-metal MOFs with OMS were investigated through tuning binding energy. This work paves the way for more robust material design for ethylene/ethane separation.

Copyright © American Institute of Chemical Engineers. All rights reserved.

569216 Design and Analysis of High-Spin Ground State Donor-Acceptor (DA) Polymers: A Computational Study

There have been a lot of efforts to design new types of organic materials having a high-spin state that can be a replacement of traditional silicon-based materials used in semiconductor, light-emitting-diodes, photovoltaics, and solar cells. Polymeric materials having alternating donor-acceptor (DA) units are under intensive study due to its solution processability, backbone flexibility, light-weight and low processing cost than silicon-based materials. For years, stable high-spin state DA polymers have been elusive to the scientific community. High-spin state organic molecules have drawn intensive interest in the field of organic chemistry as well as in materials science. Although, the unpaired electrons in high-spin molecules exert exotic properties, however, their design and characterization have confounded the scientific community due to the intrinsic instability of these molecules. Computational modelling can be used to design new types of molecules and study their properties in depth. Here, we report the design and characterization of two new donor-acceptor (DA) polymers, which have degeneracy ($\Delta E_{ST}=0$) between the singlet ($S=0$) and triplet ($S=1$) states in the neutral form. Large separation between the unpaired electrons both in singlet and triplet states with a pure diradical character ($y=1.0$) proves that polymers have a high tendency to be in a triplet ($S=1$) state for longer chain lengths. This study shows the possibility to synthesize new DA polymers, which are stable at ambient conditions, with intriguing molecular and electronic properties.

Copyright © American Institute of Chemical Engineers. All rights reserved.

570317 Stability of Ligand Protected Au Nanoclusters Under Electrochemical Reduction of CO_2

Ligand-protected gold nanoclusters (NCs) are a novel class of particles that have attracted great interest in the field of catalysis due to their atomically precise structure, high surface area-to-volume ratio, and unique electronic structure. In particular, the anionic thiolate-protected Au_{25} NC, $[Au_{25}(SR)_{18}]^{1-}$, with partially lost ligands, has been demonstrated to act as an active catalyst for the electrochemical reduction of CO_2[1, 2]. However, the stability of this and other thiolate-protected NCs after partial ligand removal remains elusive. Using density functional theory (DFT) calculations and the recently developed thermodynamic stability model (TSM)[3], we investigate the stability of $[Au_{25}(SR)_{18}]^{1-}$, $[Au_{18}SR_{14}]^{0}$, $[Au_{23}(SR)_{16}]^{1-}$, and $[Au_{28}(SR)_{20}]^{0}$ NCs when a single ligand (–R or –SR) is removed from the surface. Additionally, we examine the stability of the partially protected NCs upon the adsorption of CO_2 reduction reaction intermediates (H, CO, and COOH) on the S or Au active site generated after single –R or –SR ligand removal respectively. Our results reveal that the partially protected Au_{25} NC shows the highest stability compared to the other partially protected NCs. Additionally, we find that the presence of the COOH intermediate on the Au or S active sites stabilizes the Au_{25} NC as the removed ligand, which agrees with experimental observations on the stability of these NCs under electrochemical conditions. This study offers molecular level insights into the stability of ligand-protected gold NCs during the electrochemical reduction of CO_2 to CO.

1. S. Zhao, N. Austin, M. Li, Y. Song, S. D. House, S. Bernhard, J. C. Yang, G. Mpourmpakis and R. Jin, *ACS Catalysis*, 2018, **8**, 4996-5001.
2. N. Austin, S. Zhao, J. R. McKone, R. Jin and G. Mpourmpakis, *Catalysis Science & Technology*, 2018, **8**, 3795-3805.
3. M. G. Taylor and G. Mpourmpakis, *Nat Commun*, 2017, **8**, 15988.

Copyright © American Institute of Chemical Engineers. All rights reserved.

571876 Energy Profiles of the Ring Puckering of Cyclopentane, Methylcyclopentane and Ethylcyclopentane

The use of pseudoangles to describe five-membered ring puckering has been well established and widely used, particularly for studying ribose conformations. Several studies have investigated the relative energy of a given molecule as a function of the pseudoangle. However, most of these studies focused on a particular minimum. However, for molecules that have multiple exocyclic torsion angles along the ring (e.g. ribose) or an extended exocyclic group, a molecule can have multiple pseudoangle energetic profiles. Ethylcyclopentane is the simplest example of this, with the ethyl group adopting (+/-)-gauche and trans conformation relative to the ring. Interestingly, there has not been any experimental or computational investigations on ethylcyclopentane's ring puckering.

Herein we report all of the minima of ethylcyclopentane and its two pseudoangle rotational profiles as determined using electron correlated theories. The puckering conformations and relative energies for cyclopentane and methylcyclopentane are also reported, allowing us to investigate the influence that exocyclic methyl and ethyl groups have on the ring pucker within a uniform theoretical framework. Geometry optimizations were performed using MP2 and cc-pVTZ and aug-cc-pVTZ basis sets, with relative potential energies (rPE) computed at MP2.5 and CCSD(T) theory levels. The pseudoangle rotational profiles were computed at MP2.5/cc-pVTZ//HF/6-31G(d).

Copyright © American Institute of Chemical Engineers. All rights reserved.

575401 Effects of Coexisting Familial Mutants on Wild Type Amyloidβ_{21-30} Assembly

We present a comprehensive study using simulation and experiments to understand the co-assembly of wild type (WT) Amyloidβ (Aβ) peptide and its familial mutants. Aβ peptide has been identified to play a major role in the progression of the debilitating neurodegenerative Alzheimer's disease (AD). Aβ monomers first aggregate into soluble oligomers which further aggregates to insoluble fibrils that are deposited as senile plaques. The fibrillar structures of familial mutants of Aβ have been reported to be different than that obtained from Aβ. In heterozygous carrier patients where both WT Aβ and one of its familial mutants coexist, WT Aβ structure shows stark phenotypic difference compared to fibrils found in homozygous AD patients where only WT Aβ is present. This suggests possible differences in aggregation pathways. Here we studied the 21-30 sub-sequence of WT-Aβ ($A\beta_{21-30}$) in the presence of two of its familial mutants, E22G and E22Q. Nuclear magnetic resonance (NMR) spectroscopy and circular dichroism (CD) were used to examine both homogeneous and heterogeneous solutions of WT and familial mutants of $A\beta_{21-30}$ monomers. NMR diffusion experiments and coarse-grained molecular dynamics simulations for both pure (WT, E22G, E22Q) and mixed (WT + E22G, WT + E22Q) systems were performed. Contact maps were used to study the relative frequency of interactions between the two types of monomers in a 1:1 mixture of WT and familial mutant $A\beta_{21-30}$. The combined multiscale simulation and experimental approach allows us to understand factors which might affect the co-assembly process.

Copyright © American Institute of Chemical Engineers. All rights reserved.

576894 Studying Shape-Dependence of Structural Conformations for Coarse-Grained Thermo-Sensitive Bottle-Brush Polymer Models

Thermo-responsive polymers are a class of "smart" materials that show the ability to respond to a change in surrounding temperature. This has attracted much scientific interest for biomedical applications including drug delivery, tissue engineering, and gene delivery. In recent years, significant efforts have been made to develop coarse-grained (CG) models to facilitate the modeling of systems with larger length-scales (~m) for longer time-scales (~s). In the present study, we have employed CG model of poly(N-isopropylacrylamide) (PNIPAM) to study its bottlebrushes with different shapes. Effect of the shape, backbone length, and grafting density of the side-chains on the conformations of individual side-chain and backbone have been studied. Coarse-grained MD simulations were performed below (290 K) and above (320 K) the lower critical solution temperature (LCST) of PNIPAM in an aqueous medium. The CG MD simulation trajectories were analyzed for radius of gyration of BBPs and individual chains.

Copyright © American Institute of Chemical Engineers. All rights reserved.

578104 Computational Studies of the Oxidation and Reduction of Pd-Ag Alloy

The Pd-Ag alloy is synthesized by deposition of Pd onto Ag(111) surface at room temperature, both Pd and Ag can be oxidized by O_2 in annealing conditions. We built models of single and double PdO(101) layers on Ag(111) as well as the AgO-p(4x4), and calculated CO adsorption and H_2 oxidation on the surface by using density functional theory (DFT). The IRRAS experiments at 150 K indicated that CO adsorbs on Pd top sites rather than Pd bridge sites, CO can also adsorb on Ag oxides at higher pressure. In our calculations, we found that CO adsorbs on Ag top sites, with a much weaker binding energy than PdO(101). These results are consistent with experiments. We also found that CO adsorption on single layer PdO(101) is weaker than on double and multiple layers of PdO, which agrees with previous studies of PdO(101) on Pd(100). The AP-XPS experiments showed that pure AgO cannot be reduced by H_2 at room temperature. However AgO in the mixed oxide can be quickly reduced in the same conditions. We calculated the reduction of the oxides by H_2. We found that H_2 activation on AgO has a high barrier, which makes the reduction reaction difficult at room temperature. On the other hand, H_2 dissociates on PdO with a lower energy barrier than AgO, but the further reduction of PdO, which leads to the formation of water, has a relatively high barrier. With these calculations of reduction mechanisms, we can explain why AgO reduction can be facilitated by PdO found in experiments. H_2 can be easily dissociated on PdO, and then migrate to AgO surface. Since the water formation barrier on AgO is lower than PdO, AgO will be reduced easily.

Copyright © American Institute of Chemical Engineers. All rights reserved.

583289 Efficient Generation of Full-Atomistic Polymer Amorphous Structure Adsorbed onto Solid Surfaces by Reverse-Mapping technique

Structure generation is the beginning in investigating materials with molecular dynamics simulation, however, we have sometimes difficulties in generating geometries of molecular systems, e.g. polymer amorphous structures, micro phase separation structures, etc. Since polymer systems have long relaxation time, it is often difficult to equilibrate the systems. There are several methods for polymer amorphous structure generation utilizing Monte-Carlo (MC) based simulations such as Rotational Isomeric State (RIS)-MC and Configurational Bias (CB) MC. Another method is to consider a relation between real polymers and universal polymer models. A mapping relation between beads-spring (BS) model and full-atomistic (FA) model is proposed by Evaraers group [1]. According to the mapping relation, we have developed a scheme which constructs a FA polymer model from BS model utilizing reverse-mapping technique. In this study, the scheme is applied to generate polymer structures adsorbed on solid surfaces. In treating the BS model, Lennard-Jones Wall type potential is used to take into account the effects of the solid surface. Then FA model is successively constructed.

[1] C.Svaneborg et al., "Kermer-Grest models for universal properties of specific common polymer species", arXiv:1606.05008

Copyright © American Institute of Chemical Engineers. All rights reserved.

570545 Partial Molar Properties from Molecular Simulation Using Multiple Linear Regression

Partial molar volumes, energies, and enthalpies can be computed from molecular simulations in any open ensemble through a simple post-processing procedure that leverages fluctuations in composition, total volume, total energy, and pressure of a simulation box.

By recording the instantaneous box volume V and number of molecules of each of n species for M frames, a large M x n matrix is constructed for the composition, as well as an M x 1 vector of volumes. The 1 x n vector of partial molar volumes can then be solved simultaneously using multiple linear regression. A similar construction permits calculation of partial molar energies and enthalpies using M instantaneous measurements of the total energy and pressure of the system. Consequently, simulations need not be initiated with partial molar properties in mind; this method does not require accounting of intermolecular interactions or configurations, and imposes no restrictions on Monte Carlo moves or ensemble, so long as molecule populations fluctuate.

The method is demonstrated on two systems in the *NpT*-Gibbs and *NVT*-Gibbs ensembles: a relatively incompressible mixture of ethanol, dodecane, hexane, and water at liquid-liquid equilibrium [1] and a highly compressible natural gas condensate of methane, butane, and decane. In addition, the method is extended to predict partial molar volumes and enthalpies of adsorbed phases [2], even in rigid simulation boxes, so long as the populations of all adsorbing species are fluctuating. Different approaches for estimating statistical uncertainty are assessed, and property predictions are compared to those from equations of state.

[1] D. B. Harwood, C. J. Peters, and J. I. Siepmann, Fluid Phase Equil., 407, 269-279, (2016).

[2] R.F. DeJaco, B. Elyassi, M. D. de Mello, N. Mittal, M. Tsapatsis, and J. I. Siepmann, J. Chem. Phys., 149, 072331 (2018).

Copyright © American Institute of Chemical Engineers. All rights reserved.

565208 Improvements of the Continuous Fractional Component Monte Carlo Method

The Continuous Fractional Component Monte Carlo (CFCMC) simulation method is an expanded ensemble methodology used to solve the problem of low insertion/deletion efficiencies in open ensembles. It allows for direct calculation of the chemical potential without any post-processing. Chemical potentials can be obtained by binning of the coupling parameter λ and using the probability values p(λ=0) and p(λ=1) which require extrapolation. We show that this extrapolation leads to systematic errors when the distribution p (λ) is steep, and we propose an alternative binning scheme which significantly improves the accuracy of computed chemical potentials. Chemical potentials of coexisting gas and liquid phases for water for the temperature range 330 K to 350 K are computed using the CFCMC technique. In sharp contrast to Widom-like test particle methods, accurate chemical potentials in the liquid phase are computed using the CFCMC technique. The CFCMC simulation methodology is also applied to the reaction ensemble (serial RxMC). The key ingredient is that fractional molecules of either reactants or reaction products are present and that chemical reactions always involve fractional molecules. The advantage is that the chemical potentials of all reactants and reaction products are obtained without any post-processing. The efficiency of the algorithm can be increased significantly by applying independent biasing to the fractional molecules of reactants and reaction products. The serial Rx/CFC approach is tested for the reaction of ammonia synthesis at various temperatures and pressures. Excellent agreement was found between results obtained from serial Rx/CFC, experimental results from literature, and thermodynamic modeling using the Peng–Robinson equation of state. The efficiency of reaction trial moves is improved by a factor of 2 to 3 (depending on the system) compared to the RxMC/CFCMC formulation by Rosch and Maginn (Journal of Chemical Theory and Computation, 2011, 7, 269–279). We also combined the CFCMC simulation methodology with the original idea of Frenkel, Ciccotti, and co-workers (Chemical Physics. Letters, 1987, 136, 3) to calculate partial molar excess enthalpies and partial molar volumes of components in a single CFCMC simulation. Simulations are performed at constant mixture composition, constant temperature and pressure in the expanded version of the NPT ensemble (CFCNPT). Computation of partial molar properties in the CFCNPT ensemble does not have the drawbacks of Widom-like test particle methods, since particle insertions/removals take place in a gradual manner. Our method is applied to mixtures of NH3, N2 and H2 at chemical equilibrium. It is shown that the contribution of the partial molar enthalpies in calculating the reaction enthalpy of the Haber-Bosch process is significant at high pressures (up to 64% at a pressure of 80 MPa, relative to the reaction enthalpy at a pressure of 1 bar).

Copyright © American Institute of Chemical Engineers. All rights reserved.

569827 Improving the Efficiency of Monte Carlo Simulations of Charged Molecules in the Grand Canonical Ensemble

While ionic liquids have promising applications as industrial solvents, predicting their fluid phase properties and coexistence remains a challenge. Grand canonical Monte Carlo simulation is an effective method for such predictions, but equilibration is hampered by the apparent requirement to insert and delete neutral sets of charged molecules simultaneously in order to maintain charge neutrality. For relatively high densities and low temperatures, previously developed methods have been shown to be essential in improving equilibration by gradual insertion and deletion of these neutral sets of molecules. We introduce an expanded ensemble approach which may be used in conjunction with these existing methods to further improve efficiency. Individual charged molecules are inserted or deleted in one Monte Carlo trial rather than simultaneous transfer of neutral sets. We show how charge neutrality is maintained and show rigorous quantitative agreement between the conventional and the proposed expanded ensemble approaches, but with orders of magnitude increase in efficiency. The expanded ensemble approach is also more simple to implement than simultaneous transfer of neutral sets, and its implementation is demonstrated within the open source FEASST software (https://pages.nist.gov/feasst).

Copyright © American Institute of Chemical Engineers. All rights reserved.

576293 Mapped-Averaging Formulations for Evaluation of Singlet and Multibody Density Distributions

We show that two recently proposed methods for computing singlet and pair density distributions without histograms are particular implementations of the general mapped-averaging framework for deriving alternative ensemble averages for physical properties. We then go on to use mapped averaging develop additional expression for singlet and multibody density distributions, and examine their performance in molecular simulation in application to several systems, including those that are inhomogeneous due to an external field (e.g., particles in a pore), as well as those that form microstructure spontaneously (e.g., crystals). We consider application also to systems with discontinuous potentials, such as hard spheres, and describe the manner in which these force-based methods are applied in these cases.

Copyright © American Institute of Chemical Engineers. All rights reserved.

587479 Iterative Peptide Discovery with Active Learning and Meta-Learning of Deep Convolutional Classifiers

Often the development of novel materials is not amenable to high-throughput or purely computational screening methods. Instead, materials must be synthesized one at a time in a process and does not generate significant amounts of data. One way this method can be improved is by ensuring that each experiment provides the best improvement in both material properties and predictive modeling accuracy. In this work, we study the effectiveness of active learning, which optimizes the order of experiments, and meta learning, which transfers knowledge from one context to another, to reduce the number of experiments necessary to build a predictive model. We present a novel multi-task benchmark database of peptides designed to advance active, few-shot, and meta-learning methods for experimental design. We show results of standard active learning and meta-learning methods across these datasets to assess their ability to improve predictive models from the fewest number of experiments. We find both uncertainty minimization and query by committee to be effective active learning techniques. The meta-learning method Reptile was found to improve accuracy when only 3-5 experimental results are available, but trail normal gradient-descent based methods beyond 5 experiments.

Copyright © American Institute of Chemical Engineers. All rights reserved.

569513 Azide and Alkyne Charmm Parameterization: Resolving Dihedral Interactions That Contain Linear Moieties

Drug-like small molecules that contain azido and akynyl groups are structurally unique because they contain linear angles useful in therapeutics, protein functionalization and biomarking, and in bioconjugation reactions like "click" reactions. The click reaction has proven to be integral in drug design and in research with unnatural amino acids (uAA); however, these technologies rely heavily on molecular modeling. Improving the the accuracy of these models will lead to more accurate simulations and better rational design.

The CHARMM force field (FF) is a standard when it comes to all-atom protein and DNA modeling. CGenFF is the leading program used to generate CHARMM parameters for "drug-like" molecules that have not been explicitly modeled. However, both these FF lack appropriate parameterization to simulate any molecules that contain linear angle moieties, such as azides and alkynes, and current literature does not contain any solution for modeling these structures.

We propose a method to address this issue by (1) developing CHARMM parameters for four small molecules that contain terminal azido and alkynyl groups using the Force Field tool kit (ffTK), (2) addressing linear structure issues seen during ffTK analysis by modifying the procedure so that all CHARMM parameters can be solved, and (3) validating those results via in silico molecular dynamic (MD) simulation. We then combine these parameters with CGenFF to generate parameters for all-atom protein simulation with uAA mutations.

In this presentation we describe the process used to obtain CHARMM parameters for our selected small molecules, with a particular emphasis on determining and validating the dihedral parameters for linear moieties since these have proven very difficult to parameterize in the past. During validation, we prove that the physics of the linear terminal dihedral angles are well-represented by a large number of parameter sets without compromising molecular structure. Next, we prove the reliability of our parameter set by running MD simulations to prove our modeled structures match those found in literature and quantum theory. Protein MD next compares all-atom protein simulation with crystallographic data using a protein that contains an azide uAA mutation within its structure.

The final portion of the presentation extrapolates the all-atom parameters to generate Go-like coarse-grain uAA parameters useful for functionalized protein design. We concluded by showing uAA mutation effects on protein structure, stability and activity.

The results of this presentation resolve issues with linear moiety parameterization using the CHARMM force field, and provide an unprecedented view into how unnatural amino acids influence protein structure.

Copyright © American Institute of Chemical Engineers. All rights reserved.

564485 Large-Scale Free-Energy Calculations in Metal-Organic Framework Prototypes

The chemical and structural tunability of metal-organic frameworks (MOFs) can be exploited to alter how molecules adsorb, diffuse and react within their pore structure, making MOFs promising for applications in gas storage, chemical separations, and catalysis. However, MOF tunability also implies that there are millions of potential MOF variations, making it challenging to rapidly identify the best MOF for a given purpose. As a response to this challenge, computational high throughput screening emerged as a tool to rapidly identify promising MOF designs to be pursued by more focused (and slower) experimental efforts. This high throughput approach has already proven successful for the prediction of application-relevant adsorption properties [1] while efforts to calculate mechanical [2] and catalytic properties [3] are also starting to emerge. However, a significant obstacle that remains when progressing from the identification of a promising (in silico) MOF prototype to actually synthesizing it is the uncertainty whether the prototype is synthesizable.

Uncertainty in MOF prototype synthesizability arises due to an insufficient understanding of what makes a MOF thermodynamically and kinetically accessible. While the holistic view provided by high throughput screening of adsorption properties has provided valuable insights on what gives a MOF prototype high adsorption performance, similar insights on what makes a MOF prototype "synthesizable" have not been obtained due to the absence of large-scale efforts to calculate measures of thermodynamic stability, such as free energy, in MOF databases. This gap is perhaps due to the relative complexity of free energy calculations in porous materials. Here we present our efforts to automatize free energy calculations of MOF prototypes using the Frenkel-Ladd (FL) path method. We compare free energies calculated using the FL method to those calculated using the harmonic approximation (HA), which is a method often used to calculate crystal free energies due to the simplicity of its principles. However, the HA is a multistep process that includes the calculation of the phonon density of states (PDOS), making it a relatively involved approach not ideal for large-scale screening.

Based on our simulation automatization efforts, we find the FL method to be better suited for calculation of free-energies in large MOF databases. Furthermore, the FL method does capture the contribution of anharmonic vibrations to crystal free energy, which are important in some structures. Indeed, we show that the entropic contributions to free energy calculated using the HA method tend to match poorly with those calculated using the FL method, which could result in substantial inaccuracies when considering highly flexible MOFs. Finally, we compare the free energies of a set of MOFs calculated using three "off-the-shelf" force-fields (including the MOF-focused UFF4MOF) to gauge how much different force field parameterizations affect MOF stability predictions.

1. Gómez-Gualdrón, D. A.; Colón, Y. J.; Zhang, X.; Wang, T. C.; Chen, Y.-S.; Hupp, J. T.; Yildirim, T.; Farha, O. K.; Zhang, J.; Snurr, R. Q. Evaluating Topologically Diverse Metal–Organic Frameworks for Cryo-Adsorbed Hydrogen Storage. *Energy Environ. Sci.* **2016**, *9* (10), 3279–3289.
2. Anderson, R.; Gómez-Gualdrón, D. A. Increasing Topological Diversity during Computational "Synthesis" of Porous Crystals: How and Why. *CrystEngComm* **2019**, *21* (10), 1653–1665.
3. Rosen, A. S.; Notestein, J. M.; Snurr, R. Q. Structure–Activity Relationships That Identify Metal–Organic Framework Catalysts for Methane Activation. *ACS Catal.* **2019**, *9* (4), 3576–3587.

Copyright © American Institute of Chemical Engineers. All rights reserved.

572190 Designing Anion Exchange Membranes with Improved Ion Conductivity By Mesoscale Simulations

Understanding morphology of anion exchange membranes (AEM) and their transport properties is the key for future development of AEM fuel cells. The present work devises standard protocols for predicting the microstructure of AEM and diffusivity of hydroxide ions based on dissipative particle dynamics (DPD) simulations.[1] The protocols aim to improve the accuracy of DPD in modeling ion conductivity in AEM by adapting recently developed DPD approaches. AEM molecules are dissected into coarse grained beads with non-uniform interaction diameters based on the Connolly volume of repeating units and functional groups. Mismatch parameters are determined by activity coefficients of bead components and the calibration curves constructed by particle insertion method. Hydroxide ions which interact with water and cationic groups are modelled explicitly by the corresponding association potentials, similar to the DPD proton model in polyelectrolyte membranes.[2] The methodology is applied to study aromatic polymer PPO-TMA (polyphenylene oxide trimethylamine) and QAPS (quaternary ammonium polysulfone) with different side chain modifications, in order to avoid over-assembling of polymer backbones and to promote hydroxide transport. The method preserves the efficiency and simplicity of conventional DPD methods with improved descriptions for chemical specifics, and is expected to provide insights into the principles of new membranes design.

References:

[1] Lee, M.-T., Exploring Side Chain Designs for Enhanced Ion Conductivity of Anion Exchange Membranes by Mesoscale Simulations. *The Journal of Physical Chemistry C* **2019**, *just accepted*.

[2] Lee, M. T.; Vishnyakov, A.; Neimark, A. V., Modeling Proton Dissociation and Transfer Using Dissipative Particle Dynamics Simulation. *Journal of Chemical Theory and Computation* **2015**, *11*, 4395-4403.

Copyright © American Institute of Chemical Engineers. All rights reserved.

578122 Ultrafast Detonation of Hydrazoic Acid:Insights from Many-Body Moleculardynamics Force Fields

Understanding the behavior of energetic material under detonation is of interest for a wide range of applications in many fields, such as civil engineering and space exploration. Atomistic molecular dynamics simulations have emerged as a powerful tool to provide valuable insights into such system while circumventing the risks of experimental works. However, due to the short time scales and/or small system size accessible of quantum mechanics simulations and the lack of reliable force fields for molecular mechanics models, very little is known about the chemical evolution and states of matter of an energetic material under detonation. We have developed the Chebyshev Interaction Model for Efficient Simulation (ChIMES), a reactive force field, that retains the accuracy of density functional theory (DFT) simulation while increasing orders of magnitude in computational efficiency. In this work, we use ChIMES to study hydrazoic acid, an azide energetic material that exhibits an ultrafast detonation during a shock wave. We find that our models are able to accurately reproduce the structural properties and chemistry computed from DFT at multiple thermodynamic states. Shock compression simulations using the multiscale shock simulation technique show that our ChIMES models can capture the ultrafast chemical reactions of hydrazoic acid.
This work was performed under the auspices of the U.S. Department of Energy by Lawrence Livermore National Laboratory under Contract DE-AC52-07NA27344.

Copyright © American Institute of Chemical Engineers. All rights reserved.

557669 Computationally Aided Design of Two-Dimensional Composites As Alternatives to Metallic Copper for Catalyzing the Carbon Dioxide Reduction Reaction

In the present study, we performed the computationally aided design of a series of candidates for electrochemical catalysts that are potential alternatives to metallic copper for the carbon dioxide reduction reactions (CO_2RR) that lead to multi-carbon species. These catalysts were formulated as supported copper −nonmetal composites where small copper clusters were incorporated into the defects of the selected two-dimensional nonmetallic materials with graphene-like structures, with the aim of achieving higher efficiency and stronger selectivity compared to the pristine metallic copper. CO_2 molecules were adsorbed on the non-metallic parts of these materials and transported to the copper clusters, where they were catalytically reduced to multi-carbon molecules through carbon−carbon (C−C) coupling reactions. Using a combination of density functional theory, basic statistical mechanics, and the classical linearized Poisson −Boltzmann model [J. D. Goodpaster, A. T. Bell, and M. Head-Gordon, *J. Phys. Chem. Lett.* **2016**, 7, 1471], we constructed a chemical reaction network that includes all thermodynamically and kinetically viable surface reaction paths in the presence of the solvent, electrolyte, and applied electric potential, starting from the adsorption of CO_2 and ending with the production of two-carbon (C_2) [A. J. Garza, A. T. Bell, and M. Head-Gordon, *ACS Catal.* **2018**, *8*, 1490] and three-carbon (C_3) molecules. Based on this information, we were able to predict the efficiency and selectivity of our designed materials and to propose potential candidates for follow-up experimental measurements.

Copyright © American Institute of Chemical Engineers. All rights reserved.

565073 Tuning the Selectivity of Ethyl Disproportionation on Pt-Doped Cu(111) Surfaces: Do You Want Ethane or Ethylene?

The recent development (synthesis and characterisation) of single-atom alloys has, over the last five years, been a real breakthrough in the field of heterogeneous catalysis **[1]**. When atomically dispersed and embedded in the network of a coinage metal (Au, Ag, Cu), late transition metals show high activity and selectivity. This is related to their electronic properties, which were shown to be more similar to atoms than extended metallic surfaces **[2]**. In particular, Pt-doped Cu(111) single atom alloys (SAA), later referred to as Pt·Cu(111), are very promising as they activate C-H bonds (like pure Pt) while displaying a strong resistance to the formation of over-unsaturated hydrocarbons and ultimately coke formation (property inherited from Cu) **[3]**.

Here we study the reactivity of Ethyl as a probe molecule on both Cu(111) and Pt·Cu(111) surfaces, performing periodic Density Functional Theory calculations (VASP – optB86b-vdW) and kinetic Monte-Carlo simulations (Zacros – KMC). We consider two pathways, namely the hydrogenation to ethane and the beta C-H activation to ethylene. On Cu(111) both hydrogenation and dehydrogenation pathways show similar activation energies (Ea = 0.55 eV). Because of the contribution of the pre-exponential factors to the kinetic parameters and the difference in the order of the elementary processes, dehydrogenation happens to be 10 times faster than hydrogenation. When Pt substitutes one of the surface Cu atoms, Ethyl is more stable on top of Pt than on Cu. Moreover, Pt selectively opens the route to hydrogenation (Ea = 0.54 eV) over dehydrogenation (Ea = 0.74 eV). Pt·Cu(111) therefore becomes an interdependent bifunctional catalyst, with Cu and Pt being the active sites of dehydrogenation (production of atomic hydrogen) and hydrogenation (consumption of atomic hydrogen) respectively. Moreover, and unlike pure Cu(111), Pt is able to activate H_2 on Pt·Cu(111). With similar activation energies for hydrogenation and dehydrogenation on Pt·Cu(111), we can easily control the selectivity to either ethane or ethylene tuning the partial pressure of the atmosphere of H_2 in contact with the metallic surface.

This theoretical study is compared with surface science experiments, namely Temperature Program Desorption (TPD) spectra. The theoretical model (DFT + KMC) agrees almost quantitatively, as it is able to predict TPD peak temperatures and trends in selectivities under ultra-high vacuum conditions. Further comparison with micro-reactor studies is on-going.

This study expands our knowledge on the reactivity of Pt·Cu(111) as a single-atom alloy catalyst. It further proves the promising properties of Pt·Cu(111) as being a versatile heterogeneous catalyst with tuneable selectivity.

References:

[1] Darby M. T., Stamatakis M., Michaelides A., Sykes E. C. H. *J. Phys. Chem. Lett.*, **2018,** *9*, 5636-5646
[2] Thirumalai H., Kitchin J. R. *Topics In Catal.* **2018,** *61,* 462-474
[3] Marcinkowski M. D., Darby M. T., Liu J., Wimble J. W., Lucci F. R., Sungsik L., Michaelides A., Flytzani-Stephanopoulos M., Stamatakis M., Sykes, E. C. H. *Nature Chem.* **2018,** *10,* 325-332

Copyright © American Institute of Chemical Engineers. All rights reserved.

575408 Atomistic Simulation of Defect Formation for Quantum Information Science Applications

Defects in diamond or silicon carbide (SiC) are key material features of emerging quantum information science (QIS) applications. SiC is particularly attractive because of its ease of growth and microfabrication and integrability to existing optoelectronic devices. Its vacancy defects are potentially capable of hosting optically-active electronic spin qubits with long coherence times even at room temperature, providing the basis for quantum-based technologies. However, there is currently a limited understanding in how to reliably create the vacancy defects needed to host spin qubits. Here, we have developed a novel computational modeling approach to understand the connection among atomistic structure, material processing conditions, and electronic properties of vacancy defects in SiC. We apply advanced simulations techniques such as *ab initio* and classical molecular dynamics methods and validate our model findings with thermal annealing and photoluminescence experiments. We then go on to present molecular-level physical insights into how multiple vacancy defects are created, interact, and migrate within SiC to derive design rules for robustly constructing defects in scalable quantum materials.

Copyright © American Institute of Chemical Engineers. All rights reserved.

572983 Modeling the Sintering and Redispersion of Supported Nanoparticles By Atomistic Simulations

High performance of supported nanocatalysts are determined by their size and structures. Thus, how to reduce the size and prevent sintering of nanoparticles becomes the one of the promising route for enhancing their catalytic activities. On the other hand, recent progress of single-atom catalyst has attracted extensive attention though, the stability of single-atom catalyst is a major concern for practical usage due to sintering at high temperature. Herein, we performed a real-time atomistic simulation to reproduce the sintering process of the supported nanoparticles and single-atom catalysts combining an all-atom multiscale model and kinetic Monte Carlo simulation, which unveils a new origin of sintering and redipsersion mechanism. This work not only helps to evaluate the stability of sintering-resist nanocatalysts, but also is beneficial for the rational design of single-atom catalyst with high stability.

Copyright © American Institute of Chemical Engineers. All rights reserved.

570262 Sticking Efficiency of Silica Clusters By Molecular Dynamics

Collisions between small clusters take place in the early stages of particle formation playing a key role in environmental and industrial processes, like condensation and sintering. Such collisions lead to the formation of larger particles. However, the impinging clusters do not always stick with each other upon collision, especially if they are small. However, a sticking probability of unity is assumed typically, regardless of cluster size.

Here, molecular dynamics simulations are used to model the collision between two silica clusters consisting of 18 up to 333 atoms. The method is based on Yang et al. (2018) to obtain the sticking rate of silica clusters in the gas-phase. In these atomistic simulations, the detailed structure and surface roughness of the clusters are accounted for, contrary to the classic collision theory of hard spheres. Furthermore, criteria for successful cluster-cluster collisions are proposed based on the collision time. The effect of temperature, cluster size and orientation as well as atom velocity and collision angle on the sticking probability is investigated.

The above method predicts accurately the sticking probability and condensation rate coefficients that can be used to correct the collision frequency of agglomerates. The revised coagulation rates can be employed in greatly simplified population balance equation models coupled with fluid dynamics to facilitate the design and operation of aerosol reactors as well as in aerosol dynamics and soot formation models.

Yang, H., Goudeli, E., & Hogan Jr, C. J. (2018). *J Chem Phys*, 148, 164304.

Copyright © American Institute of Chemical Engineers. All rights reserved.

574135 Soot Particle Structure: Insights from Molecular Dynamics and Monte Carlo Simulations

Soot particle structure: Insights from molecular dynamics and Monte Carlo simulations

Kimberly Bowal[a], Jacob W. Martin[a,b], Peter Grancic[c], Markus Kraft[a,b,d]

[a]Department of Chemical Engineering and Biotechnology, University of Cambridge, West Site Philippa Fawcett Drive, Cambridge, CB3 0AS UK

[b]Cambridge Centre for Advanced Research and Education in Singapore (CARES), CREATE Tower, 1 Create Way, 138602 Singapore

[c]Institute of Soil Research, University of Natural Resources and Life Sciences, Peter-Jordan-Straße 82, 1190 Vienna

[d]School of Chemical and Biomedical Engineering, Nanyang Technological University, 62 Nanyang Drive, 637459 Singapore

Incomplete combustion produces carbonaceous particulate matter, known as soot, which negatively affects combustion devices, human health and the environment. The development of soot prediction models and mitigation strategies requires an accurate knowledge of soot particle formation and structure. Electron microscopy has revealed that mature soot particles often have a core-shell structure, in which the molecules near the centre of the particle differ in size from those in the surface layers [1]. However, it is unknown what interactions cause this partitioning and to what extent it is present in nascent particles.

In this work, the behaviour of heterogeneous polycyclic aromatic hydrocarbon (PAH) clusters was investigated using two independent molecular modelling methods to understand the structure of nascent soot particles. Clusters of up to 100 molecules containing combinations of the different sized PAHs circumcoronene, coronene, ovalene, or pyrene were evaluated. Replica exchange molecular dynamics (REMD) simulations sampled many configurations at high and low temperatures to determine stable low energy structures [2]. These results were compared with a novel Sphere Encapsulated Monte Carlo (SEMC) method, developed here to extend basin-hopping minimisation methods for use with planar aromatic molecules.

Both the REMD and SEMC simulations showed that the most stable cluster structures consist of parallel columns of stacked PAHs in a core-shell structure, in which the larger PAHs are located in the core, and the smaller PAHs are found in a surrounding shell, as seen for a representative cluster in Figure 1. This is an inverse partitioning compared to that seen in mature soot particles, proposing a unique structure for nascent soot particles and suggesting that the molecular organisation determined by intermolecular interactions is not responsible for the core-shell structure of mature soot. These results present the first simulations of soot-sized heterogeneous PAH clusters and can help further understanding of soot formation and growth processes.

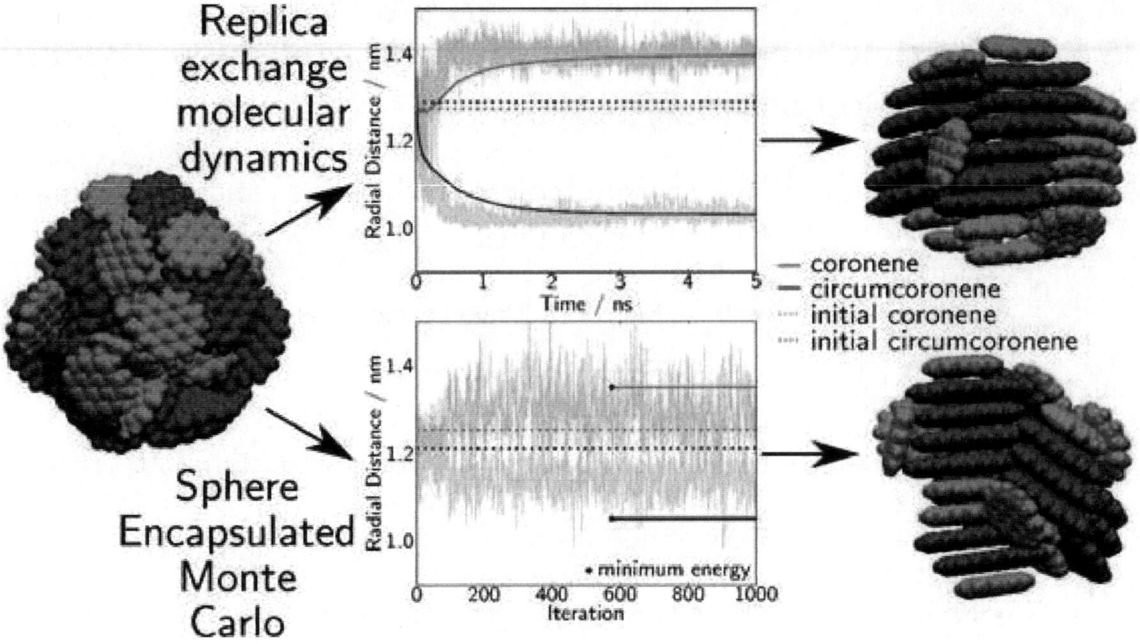

Figure 1: Radial distances of each molecule type shown for a cluster containing 16 coronene molecules (coloured red) and 16 circumcoronene molecules (coloured blue) using replica exchange molecular dynamics simulations and the Sphere Encapsulated Monte Carlo method. Initial and final cluster snapshots are also shown.

[1] T. Ishiguro, Y. Takatori, and K. Akihama. *Combust. Flame,* (1997), **108**, 231–234.

[2] K. Bowal, J.W. Martin, M. Kraft. *Carbon*, (2019), **143,** 247-256.

Copyright © American Institute of Chemical Engineers. All rights reserved.

575808 Predicting Structure and Vitrification in Toughened Epoxy Thermosets with Million-Particle Molecular Dynamics

The cross-linked network of epoxy thermosets and the nanostructure of embedded thermoplastics determine the cured material's mechanical properties. Because the bonding network and toughener phase-separation depend on the temperature history experienced during curing, it is useful to know how sensitive nanostructure and material properties are to the cure temperature history. In this work we present molecular dynamics investigations into reacting epoxy morphologies using the epoxpy package developed for HOOMD-Blue. We show how coarse-grained models of reacting epoxy thermosets simultaneously permit access to million-particle volumes and empirically validated curing dynamics, enabling the identification of models that balance computational cost with the ability to compare against experiments. We demonstrate how modulation of the curing temperature can be used to control the cured microstructure and mechanical properties. We also demonstrate the accuracy of gelation and glass transition predictions with the coarse model over a wide range of degrees-of-curing. We conclude with a discussion of how such tools can be used to predict residual stresses and inform the processing strategies used to fabricate carbon-fiber composites.

Copyright © American Institute of Chemical Engineers. All rights reserved.

575722 Molecular Dynamics Simulations of Zwitterionic Peptide Self-Assembly

Zwitterionic peptide self-assembly has been used in antifouling coating of medical devices [1], coating gold nanoparticles [2], [3], [4], biointerfaces, and peptide probed atomic force microscopy [5]. The sequence and number of charged amino acids in zwitterionic peptides determine their structural properties [6], [7]. Because amino acids that compose peptides are also found in nature, we can look for insight from nature. The surface of proteins is commonly populated by charged amino acids; they make up about 40% of human protein surfaces [8]. The most common are lysine (E) and glutamic acid (K), which are capable of electrostatically interacting with water as well as each other. Their ionic hydration creates a strong solvation shell around them, as opposed to arginine (R) which interacts with E more than water [1], [2]. Based on their strong hydration and other evidence, E and K provide a resistance of non-specific binding and that may be the reason for their common occurrence on protein surfaces [4]. There are still open questions about the interactions between these charged groups, especially regarding the seemingly small chemical difference between aspartic acid (D) and E which has vastly different occurrences on protein surfaces. In this work, we simulate large multivalent zwitterionic peptide chains, (EK)6, (ED)6, (RK)6, and (RD)6, in explicit water using molecular dynamics with the enhanced sampling technique called parallel tempering well-tempered ensemble (PT-WTE) [9]. We examined the interplay between salt bridging, hydration, peptide structure, and multivalent interactions. We used machine learning to isolate free energy basins and then compute a reaction network model (Markov state model) describing the kinetic and thermodynamic relationship between hydration and salt bridging. Our network model gives an interpretable understanding and reaction rates for the exchange of water, salt bridging, and the complexities of multiple interactions. Future work will introduce the complexity of monovalent and bivalent salts to see their role in these reaction network models.

References

[1] – E. R. L. Brisson, Z. Xiao, and L. A. Connal. Aust. J. Chem. 69.7 (2016), p. 705.

[2] – H. Ye et al. ACS Appl. Mater. Interfaces 7.40 (2015), pp. 22448–22457.

[3] – A. Parnsubsakul, S. Oaew, and W. Surareungchai. Nanoscale 10.12 (2018), pp. 5466–5473.

[4] – A. K. Nowinski et al. Langmuir 30.7 (2014), pp. 1864–1870.

[5] – R. Chang et al. Polym. J. 50.8 (2018), pp. 563–571

[6] – R. F. Sommese et al. Protein. Sci. 19.10 (2010), pp. 2001–2005.

[7] – A. K. Nowinski et al. J. Am. Chem. Soc. 134.13 (2012). PMID: 22401132, pp. 6000–6005.

[8] – A. D. White et al. Chem. Sci. 3.12 (2012), pp. 3488–3494.

[9] - M. Bonomi and M. Parrinello. Phys. Rev. Lett. 104.19 (2010), pp. 1–4.

Copyright © American Institute of Chemical Engineers. All rights reserved.

569385 A Computational Study on Ionic Liquid Cation-Cytochrome P-450 Complexes Using QM/MM Calculations to Provide Structural Insights into Their Binding

Ionic liquids have been projected to be environmentally benign owing to their characteristics like inherent low vapor pressure and flammability. Their extremely low vapor pressures translate to the fact that they have negligible role in air emissions as compared to conventional industrial solvents. These liquids can also be flexibly designed in order to tune their physical properties. Out of the several already existing ionic liquid classes, imidazolium-based ionic liquids have been one of the most successfully utilized in diverse applications. Suitable variants of these types of ionic liquids have been applied in many process in the chemical industry. Though efficacious, investigations by various experimentalists have raised questions on their environmental degradability. Thus, including rational design into their synthesis becomes imperative. Despite a lot of experimental efforts in this direction, molecular level details have not been explored in detail computationally. The present work aims to provide physical insight into the phenomena of ionic liquid biodegradability to aid in the rational design of these solvents.

The Cytochrome P-450 enzymatic superfamily has been identified and widely studied for their role in oxidation of a wide variety of molecules in both aerobic and anaerobic environments. Former experimental work suggests oxygen insertion through hydroxylation at the terminal position of alkyl chains in the cation to aid in their ability to biodegrade. Thus, it was deemed necessary to capture the effects of the P-450 protein on imidazolium-based ($[C_n\text{mim}]^+$) cations to develop a computational framework for their biodegradability. In the present work, the cations have been included in the model as substrates responsible for the reorganization of the binding pocket of cytochrome P-450. To capture the effects of substrate inclusion, the system was subjected to QM/MM calculations. The active site consisting of the heme molecule and the cation was treated at a quantum mechanical level (QM region) while the neighboring residues were treated at a classical level (MM region). Docking calculations were performed for generating suitable starting conformations for the above treatment. In the docking process, cations were inserted into the pocket varying the 1-n-alkyl chain on the cation progressively along the homologous series (n =2,4,6,8,10). Binding energy and other relevant analyses were performed on the resulting structures to provide mechanistic insights into the system. Also, the interaction strength between the heme and substrate molecules were also probed along with the stabilization of the QM region by the surrounding protein.

Copyright © American Institute of Chemical Engineers. All rights reserved.

583199 Introduction

A concise overview of the 50th Anniversary of Computer Aides for Chemical Engineering (CACHE) Conference on “The Future of Cyber-Assisted Chemical Engineering Education,” held in Breckenridge, Colorado, in July 2019 will be provided.

Dissemination plans for the conference outcomes include publication of several articles in the flagship magazine of AIChE, Chemical Engineering Progress, and a special issue of Chemical Engineering Education.

Copyright © American Institute of Chemical Engineers. All rights reserved.

583200 CACHE History

The proposal founding the Computer Aids for Chemical Engineering Committee was submitted to the NSF on July 14, 1969 and first official meeting of the CACHE organization was held on November 18, 1969 in Washington DC. From its inception, CACHE has promoted the development and distribution of technology-based materials and software in chemical engineering education through projects applying computational chemical engineering, sponsoring conferences, recognizing outstanding contributions, and providing leadership in chemical engineering education. The purpose of this presentation is to review the accomplishments of CACHE over this 50-year period in supporting educational and research innovations.

Copyright © American Institute of Chemical Engineers. All rights reserved.

583201 CACHE Initiatives: Systems Biology

In 2005, CACHE organized a new conference series entitled FOSBE (Foundations of Systems Biology in Engineering). The goal was to create a focused meeting on the topic of systems biology, as it relates to engineering methods and approaches, with a combined emphasis on: (i) applied problems from the pharmaceutical industry, (ii) research methodologies, and (iii) challenges in education and training.

The types of questions that were intended to be addressed by the conference included:
• How can one use systems biology and genomic data to analyze, interpret, and predict the relation between an organism's genotype and its phenotype?
• What are the best methods to combine data from hypothesis-driven research with data from high-throughput studies to create models of cells, communities of cells, and entire organisms? How can one drive iteration and innovation in this model building process?
• What is an appropriate computational infrastructure for maximizing the mining of bioinformatic data? What are the data format and databasing challenges?
• How do the complex network structures constrain intracellular signaling processing?
• What are the promising techniques for estimating model parameters from high throughput data records? How can identification methods be used to drive effective design of experiments?
• How can biological domain knowledge be combined with systems engineering methods to yield model reduction methodologies that capture essential features of biological regulation across multiple scales of time and space?
• How do we educate, train and develop the systems biologists of the future?

In this talk, I will report on the state of the FOSEB conference series, its unique position among other symposia and conferences, its collaborations with other professional societies, and the plans for the future.

Copyright © American Institute of Chemical Engineers. All rights reserved.

583202 CACHE Initiatives: Molecular Modeling

The molecular modeling initiative in CACHE began in 1995 as its first foray into computational-based chemical engineering education beyond process design, operations and control. Molecular modeling has now been firmly established as one of the core competencies of CACHE, with around one quarter of the Board of Trustees representing this effort. In this talk, we will review the activities of CACHE in the field of molecular modeling, which includes educational initiatives and the establishment of the successful Foundations of Molecular Modeling and Simulation (FOMMS) conference series (fomms.org) that has been held every three years, beginning in 2000.

Copyright © American Institute of Chemical Engineers. All rights reserved.

583203 CACHE Initiatives: Process Systems

This presentation describes CACHE initiatives in the area of process systems. CACHE supports the teaching of chemical engineering courses by managing webpages that link to resources that include syllabi, course schedules, computer-aided tools, interactive simulations, screencasts, concept questions, textbook information, useful links, and course notes. The resources relevant to the teaching of process systems include computer-aided educational tools developed by CACHE, such as sustainable manufacturing educational modules.

CACHE organizes conferences on progress and future directions in process systems, which include Chemical Process Control, Foundations of Computer Aided Process Operations, and Foundations of Computer-Aided Process Design. Activities and results are summarized for a newly founded conference on Foundations of Process Analytics and Machine Learning.

Copyright © American Institute of Chemical Engineers. All rights reserved.

583204 Beyond A -> B: Computational Approaches for Education in Reaction Engineering and Kinetics of Complex Systems

Chemical reaction engineering (CRE) is a hallmark of the education of chemical engineers (ChE). It is often cited as a distinguishing feature a ChE compared to other engineers from other disciplines due to its clear marriage of chemistry and engineering. This presentation will focus on cyber-enabled tools that can be easily integrated into the undergraduate curriculum to introduce students to more complex reaction systems governing archetypal chemical processes, where an understanding of reactions well beyond A ® B is essential to guide process design and improvement of both novel and mature technologies. Automated tools for reaction network generation and approaches for automated model building will be discussed. Alternative solution methods for complex reaction systems based on stochastic simulations will be introduced, and software packages that are accessible for undergraduates to carry out kinetic Monte Carlo (kMC) simulations will be outlined. The inclusion of kMC as a core element of a CRE course will be motivated through examples of phenomena that are difficult to capture through analytical rate laws and how intuitive event-based simulations are to students.

Copyright © American Institute of Chemical Engineers. All rights reserved.

583205 Graduate Research & Training in Molecular Data Science Through a University Wide Approach to Integrative Data Science Education

All areas of science and engineering have been the beneficiaries of a data-deluge. Chemical Engineering is at the forefront of this wave as a result of the commoditization of tools such as robotic instrumentation for molecular synthesis and characterization, ever increasing computational resources for simulation and design, high throughput experimental methods including gene sequencers, a proliferation of inexpensive and networked sensors and mixed modality data sources. To be competitive in the changing landscape of industry and academia, the next generation of chemical engineers will need to know how to effectively use data science tools such as statistical and machine learning, scalable data management and visualization techniques to communicate their findings effectively to stakeholders and peers.

The University of Washington (UW) has embraced data science in a campus wide initiative and the Chemical Engineering (ChemE) department at UW has led the effort to integrate highly contextualized molecular data science instruction at the graduate level, for both MS and PhD. ChemE @ UW offers two transcriptable options to graduate students, Advanced Data Science & Data Science. Students who participate in these options get a graduate degree in Chemical Engineering with a transcripted recognition of their participation in the program. The former is designed for chemical engineers who want to design and create new data science algorithms and tools, e.g. new machine learning methods for sparse molecular descriptor spaces, while the latter is intended for data science tool users who want to be confident in their selection of methods, e.g. choosing an effective neural network architecture for a predictive model of blood brain barrier permeability.

This presentation will review the structure of ChemE @ UW's graduate data science options from its inception as a result of two NSF training grants and in the context of the UW campus wide investment in data science education, research and training. It will highlight the NSF NRT Data Intensive Research Enabling Clean Technology (DIRECT) program which has been the primary vehicle for sustainable graduate molecular data science education in the department that includes 1) new graduate coursework – a sequence of flipped-classroom style courses that teach Data Science skills in a manner contextualized to molecules and 2) A capstone project that uses project based learning to practice and apply data science skills to challenging real world problems (in conjunction with external partners) in a team-based setting. It will close with outcomes from early cohorts that participated in the transcriptable options.

Copyright © American Institute of Chemical Engineers. All rights reserved.

583206 Survey: Computing in Chemical Engineering

In 2003 the CACHE (Computer Aids for Chemical Engineering) Corporation carried out a survey of recent graduates in chemical engineering from four universities to determine how that group used (or did not use) computing in performing their job. Since that time, there have been considerable changes in the field of information technology, hence it is appropriate now to assess how the general conclusions reached fifteen years ago have changed. It is also valuable to have such information in order to consider changes in the curriculum, especially with respect to the use of and instruction about computing. Therefore we are conducting a similar survey to the one used in 1997 and 2003.

This survey is in two parts. One part is being emailed to chemical engineering faculty to assess what computing tools are currently being taught and used in the chemical engineering curriculum at universities. A second survey is being emailed to engineers in the chemical industry to assess computing in industry.

Copyright © American Institute of Chemical Engineers. All rights reserved.

580996 Molecular Recognition of Ice By Proteins: From Ice Nucleation to Antifreeze

Bacteria, insects and fish that thrive at subfreezing temperatures produce proteins that bind to ice and manage its formation and growth. Ice binding proteins include antifreeze proteins and ice-nucleating proteins. The latter are the most efficient ice nucleators found in Nature. Many questions remain on how do these proteins recognize or nucleate ice, what drives their selectivity and binding to ice, and how the size and aggregation of the proteins modulate their function. In this presentation, I will discuss our recent work addressing these questions using molecular simulations and theory,[1-7] with particular focus on elucidating what intermolecular interactions and chemical motifs make these proteins so efficient at their function, to resolve the apparent paradox that the same structures can promote and prevent ice formation, and to draw general principles that can be used for the design of synthetic alternatives for control of ice formation and recrystallization.

References:

1. Hydrogen-Bonding and Hydrophobic Groups Contribute Equally to the Binding of Hyperactive Antifreeze and Ice Nucleating Proteins to Ice, A. Hudait, Y. Qiu, N. Odendahl, and V. Molinero, J. Am. Chem. Soc. 141 (2019) 7887-7898

2. How Size and Aggregation of Ice-binding Proteins Control their Ice Nucleation Efficiency, Y. Qiu, A. Hudait, and V. Molinero, J. Am. Chem. Soc. 141 (2019) 7439-7452

3. Preordering of water is not needed for ice recognition by hyperactive antifreeze proteins, A. Hudait, D. R. Moberg, Y. Qiu, N. Odendahl, F.Paesani, and V. Molinero, Proc. Nat. Acad. Sci. 115 (2018) 8266-8271

4. Ice-nucleating and Antifreeze Proteins Recognize Ice Through a Diversity of Anchored Clathrate and Ice-Like Motifs A. Hudait, N. Odendahl, Y. Qiu, F. Paesani, and V. Molinero, J. Am. Chem. Soc. 140 (2018) 4905–4912

5. What Controls the Limit of Supercooling and Superheating of Pinned Ice Surfaces? P. M. Naullage, Y. Qiu, and V. Molinero, J. Chem. Phys. Lett. 9 (2018), 1712–1720

6. Antifreeze Glycoproteins Bind Reversibly to Ice via Hydrophobic Groups K. Mochizuki and V. Molinero, J. Am. Chem. Soc. 140 (2018), 4803–4811

7. Molecular recognition of ice by flexible molecules P. M. Naullage, L. Lupi, and V. Molinero, J. Phys. Chem. C 121 (2017) 26949–26957

Copyright © American Institute of Chemical Engineers. All rights reserved.

580998 Too Small Is Never Too Small: Utilizing Advanced Path Sampling to Probe the Kinetics and Mechanism of Rare Events, from Atmospheric Ice Nucleation to Membrane Transport

More often than not, scientists are challenged with the daunting problem of measuring or computing astronomically small quantities that are related to the occurrence of rare events. A phenomenon is called a rare event when the amount of time that elapses before its occurrence is orders of magnitude longer than the time needed for its completion. Rare events are ubiquitous in nature, and span a wide range of phenomena such as earthquakes, telecommunication and power grid failures, protein folding, genetic mutations, and crystallization. Capturing the statistical nature of such events is key in many applications, including materials synthesis, climate modeling, bioengineering and medicine. Unfortunately, achieving this with conventional experiments or simulations is inefficient at best as the waiting times for observing a single rare event can surpass the experimentally or computationally accessible timescales by several orders of magnitude. This becomes an almost impossible undertaking when the rate of occurrence of a rare event is astronomically small. Under such circumstances, specialized sampling techniques are necessary for capturing the statistical features of the corresponding rare event.

This presentation will be dedicated to showcasing our work on characterizing rare events using an advance sampling technique known as forward flux sampling (FFS) [1], alongside a brief discussion of a novel variant of FFS recently developed in my group suited for studying aggregation phenomena such as nucleation. I will, in particular, discuss our application of FFS to probe the kinetics and microscopic mechanism of two scientifically and practically important rare events, namely ice nucleation under atmospherically relevant conditions [3-7], and ion and solute transport through semipermeable nanoporous membranes [8]. An emphasis will be put on not only the ability of FFS to characterize their kinetics, but also its power to provide molecular insights into the underlying mechanisms of these processes.

[1] Allen et al., J. Chem. Phys. 124: 194111 (2006).
[2] Haji-Akbari, J. Chem. Phys. 149: 072303 (2018).
[3] Haji-Akbari, Debenedetti, PNAS, 112: 10582 (2015).
[4] Haji-Akbari, et al., Phys. Chem. Chem. Phys., 16: 25916 (2014).
[5] Haji-Akbari, Debenedetti, PNAS, 114: 3316 (2017).
[6] Gianetti, et al., Phys. Chem. Chem. Phys., 18: 4102 (2016).
[7] Hussain, et al., Under preparation.
[8] Malmir, et al., Submitted [arXiv:1906.00853].

Copyright © American Institute of Chemical Engineers. All rights reserved.

580997 Combining Density Functional Theory Calculations with Machine Learning for Studying Complex Chemistries on Surfaces

Global climate concerns, depletion of fossil resources, and increased demand for second-generation renewable fuels and chemicals have stimulated research in utilizing biomass derivatives. However, the complexity of the reaction network of heterogeneously catalyzed biomass derivatives, which often involve hundreds of intermediates and transition states, greatly hinders the computational study and design of new catalysts for utilizing biomass. To reduce the large cost of computing different adsorptions energies and transition state energies on various active site models, linear scaling relationships have been developed for adsorption and transition state energy prediction, such as transition state scaling (TSS) and BEP. In these linear models, only a few descriptor values are needed to calculate the energies of various intermediate species and transition states. However, these linear relationships are usually designed for small molecules and the effectiveness and applicability to more complex biomass-based chemistries is unknown. Therefore, we systematically studied the performance of linear scaling versus advanced non-linear machine learning (ML) models such as support vector regression (SVR), kernel ridge regression (KRR) and Gaussian process (GP) modeling in the hydrodeoxygenation (HDO) of organic acids. We first trained our models using data from the HDO of propanoic acid (Pac) to obtain best models and descriptors that can be applied for larger reactants, i.e., the HDO of succinic acid (SUCC). We show that with the help of ML models and proper selection of descriptors, we are able to extrapolate between molecules with a different number of carbon and oxygen atoms.

Copyright © American Institute of Chemical Engineers. All rights reserved.

580999 Modeling Nucleation of Organic Molecules with the String Method

Understanding and controlling the nucleation of molecular crystals is a major technical challenge with implications in the manufacture of food, specialty chemicals, construction materials, optical devices, and drugs, among others. The size of critical nuclei, coupled with the stochastic nature of nucleation and the time scales involved, present challenges for both experimental and simulation studies. In this talk, we will present an overview of systems we have studied using the String Method in Collective Variables, in order to understand the nucleation of organic molecules, both from a supercooled melt and from solution. Our approach allows for simulating nucleation from solution at constant supersaturation, by maintaining constant the chemical potentials of the solute, solvent, or both, eliminating solute depletion artifacts in the free energy profiles.

Copyright © American Institute of Chemical Engineers. All rights reserved.

577481 Rigid Seeding: A Computationally Efficient Method to Study Heterogeneous Nucleation in Molecular Simulations

Nucleation is the primary step towards crystallization and is significantly difficult to study in both experiments and simulations. The length and timescale resolution required to observe nucleation is difficult to access in experiments. In simulations, although the length and timescales are well-suited, the sampling required to get statistically relevant nucleation events is computationally prohibitive. Thus, novel methods are required that can enable us to study nucleation in molecular simulations in an efficient manner.

We present a computationally efficient method called Rigid Seeding (RSeed) to study heterogeneous nucleation. Heterogeneous nucleation refers to nucleation that occurs in the presence of a foreign particle (i.e. a surface). This methodology builds on the seeding method but enables accounting for the role of surface in determining the crystal structure that is nucleated on the surface. Briefly, in the RSeed method we place a rigid crystal seed on the surface and allow it to relax. During this process, the crystal seed is able to translate and rotate which enables the seed to identify the favorable seed-surface configuration.

We demonstrate the method through studies of heterogeneous ice nucleation. It has been shown that various polymorphs if ice such as hexagonal and cubic can form on surfaces and there is no obvious method to predict this. We study two surfaces: a flexible self-assembled monolayer (SAM) with COOH terminal group and a modified mineral kaolinite surface (kao_m). Using RSeed and all atom simulation we determine the ice nucleating behavior of each surface. We find that RSeed method is able to identify the ice polymorph, plane, and rotation of the ice structure that grows on both surfaces. The results are in good agreement with the nucleation trajectories from straightforward MD. We also use the RSeed method to extract the critical nucleus size on kao_m surface and calculate the rate of nucleation by extending the classical nucleation theory (CNT) to heterogeneous case. The rate calculated using straightforward MD is within the error bars of the rate predicted by heterogeneous CNT. This technique enables us to study ice nucleation on a broad range of surfaces and thus, elucidate the correlations between surface properties and its ice nucleating efficiency. RSeed method is broadly applicable and can be used to study heterogeneous nucleation in a variety of systems.

Copyright © American Institute of Chemical Engineers. All rights reserved.

562789 Theoretical and Computational Investigation of the Impact of Free Interfaces on Ice Nucleation

Vapor-liquid interfaces are known to accelerate heterogeneous ice nucleation at their vicinity, in a phenomenon known as contact freezing [1]. Their impact on homogeneous nucleation, however, is more ambiguous, as the experimental evidence for surface freezing [2], or enhancement of homogeneous nucleation at free interfaces, is not conclusive. Computational investigations of surface freezing also predict conflicting qualitative behaviors for different water models [3-4]. It has, however, been suggested that contact and surface freezing are closely related phenomena because if a free interface is capable of enhancing heterogeneous nucleation at its vicinity, it must also have the same effect on homogeneous nucleation. In this work, we use our newly developed jumpy forward-flux sampling (jFFS) algorithm [5] and molecular dynamics (MD) simulations to explore this relationship, and observe that tetrahedral liquids undergoing surface freezing will also exhibit accelerated heterogenous nucleation kinetics when the crystal-nucleating agent is sufficiently close to the free interface. By analyzing nucleation pathways, we identify a distinct "funneling" mechanism for situations in which heterogeneous nucleation is considerably accelerated. We are able to quantitatively explain these results by developing a simple CNT-based theoretical description of heterogeneous nucleation close to a free-surface.

1- A. J. Durant, Geophys Res Lett, 32, L20814 (2005).
2- A. Tabazadeh, PNAS, 99, 15873 (2002).
3- A. Haji-Akbari, PCCP, 16, 25916 (2014).
4- A. Haji-Akbari, PNAS, 114, 3316 (2017).
5- A. Haji-Akbari, JCP, 149, 072303 (2018).

Copyright © American Institute of Chemical Engineers. All rights reserved.

576152 Evaluation of the Interface Potential Via a Force-Based Virtual Box Approach within the Isothermal-Isobaric Ensemble

The wetting properties of fluids on solid surfaces play a significant role in industrial applications, such as material synthesis and design. We have pursued a so-called interface potential approach, wherein one considers the growth of a thin fluid film from a solid surface. Two versions of the approach are employed: spreading and drying. The spreading (drying) interface potential provides the surface excess free energy as a function of the thickness of a thin liquid (vapor) film adjacent to the solid surface and in a mother vapor (liquid) phase. These free energy curves provide insight into the qualitative aspects of the wetting behavior and quantitative measures of wetting properties. Our group previously developed a means to compute these interface potentials via Transition Matrix Monte Carlo (TMMC) simulation within the grand canonical ensemble. In this presentation, we describe a new means to compute the interface potentials within an isothermal-isobar ensemble. The approach employs a force-based strategy to compute the free energy. More specifically, the average force the fluid imparts on the surface is calculated at multiple film thicknesses, and these data are integrated to determine the interface potential. Within the spreading method, one samples configurations with a highly elongated simulation box. As such, we leverage a virtual box strategy in which one replaces the bulk-like interior of the pore with a virtual box that is handled analytically. A combined Molecular Dynamics and Monte Carlo simulation algorithm is used to improve sampling and accuracy. We present results for (1) a Lennard-Jones fluid at an atomistic FCC substrate at different temperatures and substrate strengths and (2) SPC/E water at a structureless 9-3 surface.

Copyright © American Institute of Chemical Engineers. All rights reserved.

563138 Free Energy Calculations Along Entropic Pathways: Application to Capillary Phase Transitions

We analyze the capillary condensation and evaporation processes in fluids confined in a cylindrical nanopore. For this purpose, we define the entropy of the adsorbed fluid as a reaction coordinate and determine the free energy associated with both processes along entropic pathways. For capillary condensation, we identify a complex free energy profile resulting from the multi-stage nature of this phenomenon. We find capillary condensation to proceed through the nucleation of a liquid bridge across the nanopore, followed by its expansion throughout the pore to give rise to the stable phase of high density. In the case of capillary evaporation, the free energy profile along the entropy pathway also exhibits different regimes, corresponding to the initial destabilization of the layered structure of the fluid followed by the formation, and subsequent expansion, of a bubble across the nanopore.

Copyright © American Institute of Chemical Engineers. All rights reserved.

568489 Determining the Predictability of Clusters Expansions By Quantifying Their Associated Errors

Quantifying the lateral interactions between adsorbates on the surface of a catalyst using a first-principles-based lattice gas Hamiltonian provides an opportunity to obtain *ab initio* insights into large-scale phenomena and presents a path forward to bridging the atomistic and meso-scale length and time scales of a heterogenous catalytic reactions. However, it has recently been shown that supercell volume and geometric relaxations—often critical to capturing the true chemistry of a system—can cause anything from poor convergence of the associated cluster expansion (CE) to outright CE failure [1-3]. Indeed, when the positions of the surface metal atoms deviate significantly from those of the bulk, a cluster expansion stops becoming useful unless many-body terms are added to a lattice gas Hamiltonian, which in turn affects its convergence. We are therefore motivated to determine how relaxations quantitatively affect cluster expansions. To do this, we have developed a robust method of quantifying errors in the effective cluster interactions (ECIs) of CEs with minimal statistical assumptions about the underlying data and model. The errors are found by direct evaluation of the ECI fluctuations occurring during the calculation of the leave-multiple-out cross-validation (LMO CV) score [4].

ECI errors were determined for two separate systems (Figure 1): O/Fe(100) with a fully relaxed lattice as well as O/Fe(100) with an idealized, fixed lattice (data for the former taken from Bray et al.[5]). The optimum CEs for these systems were found with the newly developed *ab initio* Mean-field Augmented Lattice Gas Modeling (AMALGM) code, and Figure 1 shows the results for the relaxed (Figure 1A and 1B) and fixed (Figure 1C and 1D) lattice CEs. Both systems were evaluated using the optimum CEs for the relaxed (Figure 1A and 1C) and fixed (Figure 1B and 1D) systems to fairly compare the two. Our results show that relaxations do indeed result in more uncertain CEs: regardless of which CE is used, the LMO CV score of the fixed system is consistently two to three times smaller than that for the relaxed system—and our method can quantify this uncertainty. For instance, the error in the adsorption energy value (the ECI for Cluster ID #1) is four times greater for the relaxed system (Figure 1A) than for the fixed system (Figure 1C) even when using the optimum CE for the relaxed system. Many ECIs in both systems have ECI errors that are larger than the ECI values themselves; thus, our method reveals when it is difficult to ascertain whether a particular ECI should be regarded as repulsive or attractive and when an ECI value of 0 meV would potentially be more appropriate. This suggests that we may be able to use these errors as a CE “filter”, allowing only CEs with significant ECIs to be considered during the CE optimization procedure. A particularly important observation based on our evaluated errors here is that the ECIs between the two systems are often not statistically differentiable, meaning the fixed system CE (Figure 1C) could be used in the place of the relaxed system CE (Figure 1A) with potentially little negative consequences. While this does not mean that CEs are unable to capture relaxations outright, it does highlight the importance of being able to quantify the errors inherent within a CE, and our work here provides the tools necessary to make that evaluation.

[1] J. M. Sanchez, Journal of Phase Equilibria and Diffusion 38 (2017) 238-251.

[2] J. M. Sanchez, Phys. Rev. B 95 (2017) 216202.

[3] A. H. Nguyen, C. W. Rosenbrock, C. S. Reese, G. L. W. Hart, Phys. Rev. B 96 (2017) 014107.

[4] K. Baumann, Trends Anal. Chem. 22 (2003) 395-406.

[5] J. Bray, G. Collinge, C. Stampfl, Y. Wang, J.-S. McEwen, Top. Catal. 61 (2018) 763-775.

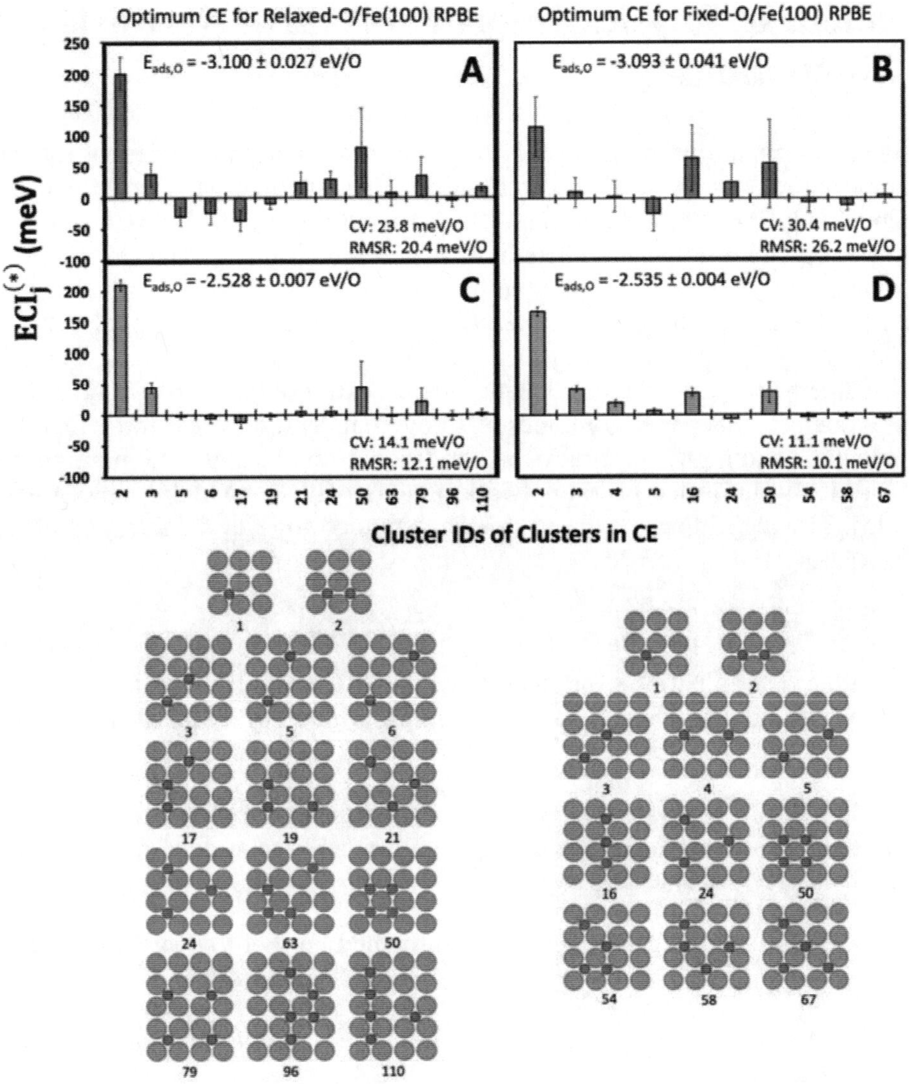

Figure 1. The ECIs found for the two optimum LG CEs fitted to the adsorption energies of the O/Fe(100) system using the RPBE exchange correlation functional and where Fe(100) surface and the oxygen adatoms in panels (A-B) were allowed to fully relax during optimization (color coded red), and where the Fe(100) surface was kept fixed in its optimum clean surface configuration and the oxygen adatoms were kept fixed in their isolated-oxygen optimum position on this surface in panels (C-D) (color coded blue). The corresponding arrangement of the oxygen adatoms for the given cluster IDs is shown below where the brown circles are Fe atoms and the red circles are O adatoms. The LMO CV score and RMSR are shown as an inset and the errors determined using our method are shown as error bars. Note that the ECIs for Cluster ID #1 (the adsorption energy of an isolated O atom) and its associated error are given in text as an inset due to magnitude discrepancy.

Copyright © American Institute of Chemical Engineers. All rights reserved.

574451 Development of Physics-Informed Neural Network Potentials for Molecular Simulations

In molecular dynamics (MD) simulations, prediction of forces acting on atoms are dependent on force fields. In the past decades, numerous physically meaningful force field equations were proposed, such as the Lennard-Jones potential [1], CHARMM force field [2] and bond order potential (BOP) [3]. However, these models which have a few adjustable parameters have been limited in accuracy. In recent years machine learning has gained a lot of attention. For example the artificial neural network (NN) has been employed to predict potential energies and interatomic forces with high accuracy through interpolation among energies from the density functional theory (DFT) database [4]. However, most NN potentials lack enough physics in their mathematical construction leading to bad extrapolation ability and poor transferability to unseen systems. This paper combines NN potentials and classical force fields to improve transferability of NN potentials and higher accuracy than traditional force fields. The pure gold system is being studied with the physically-informed neural network potential (PINN) which has been demonstrated to give accurate results [5]. The algorithm is currently being extended to relevant catalytic systems, including water/gold interfaces.

References:

[1] Jones, J. E. (1924). On the determination of molecular fields.—II. From the equation of state of a gas. *Proceedings of the Royal Society of London. Series A, Containing Papers of a Mathematical and Physical Character*, *106*(738), 463-477.

[2] Brooks, B. R., Bruccoleri, R. E., Olafson, B. D., States, D. J., Swaminathan, S. A., & Karplus, M. (1983). CHARMM: a program for macromolecular energy, minimization, and dynamics calculations. *Journal of computational chemistry*, *4*(2), 187-217.

[3] Tersoff, J. (1988). New empirical approach for the structure and energy of covalent systems. *Physical Review B*, *37*(12), 6991.

[4] Behler, J., & Parrinello, M. (2007). Generalized neural-network representation of high-dimensional potential-energy surfaces. *Physical review letters*, *98*(14), 146401.

[5] Pun, G. P., Batra, R., Ramprasad, R., & Mishin, Y. (2018). Physically-informed artificial neural networks for atomistic modeling of materials. *arXiv preprint arXiv:1808.01696*.

Copyright © American Institute of Chemical Engineers. All rights reserved.

570332 Molecular Dynamics Simulations to Study the Tribological Behavior of Gel-Forming Polymers for Cement-Based 3D Printing Applications

Molecular dynamics simulations to study the tribological behavior of gel-forming polymers for cement-based 3D printing applications

Hajar Taheri Afarani[1], Babajide Y Onanuga[1], Venkat Padmanabahan[1], and Joseph J. Biernacki[1]

[1]Department of Chemical Engineering, Tennessee Technological University, Cookeville, Tennessee

Additive manufacturing (AM), also known as 3D printing (3DP), offers new horizons to the construction industry. 3DP in construction promises reduced construction time, elimination of form work and hence greatly reduced cost, free-form shape and design and reduced waste. To make portland cement-based pastes printable, additional ingredients such as gel-forming polymers are required. However, not all gel-forming polymers provide the necessary properties for effective 3D printing. This work aims to address key questions, including: What molecular features are critical? What forms of water-polymer-cement interactions are desirable, and how do gel-forming polymers respond with cement particles under shear? To answer these and other questions, molecular dynamics simulations (MD) of various hydrogels with the (040) surface of the tricalcium silicate crystal (the primary component of portland cement) were performed. The distribution of water molecules between two cement surfaces at different concentrations of gel-forming polymers was characterized and the effect of shear rate on tribological properties computationally determined. Finally, the effect of polymer chemistry on the printability of cement pastes was also investigated.

Copyright © American Institute of Chemical Engineers. All rights reserved.

561652 Nano-Thermodynamic Model Based on Distribution Coefficients for Understanding Segregation in Metal Alloy Nano Particles

Multi-metal alloy nano particles (NPs) offer new avenues for exploration and design of nano scale-properties, e.g., catalytic, electronic and optical, by virtue of their tunable composition. Unfortunately, a method that can aid such exploration by accurately predicting the size, shape and composition dependent elemental distribution associated with nano materials is crucially missing. A nano-thermodynamic model based on distribution coefficient (Δ) is introduced to fill this gap. Δ is employed to predict surface segregation in NPs as a function of the NP size and composition. Interestingly, we find Δ to be independent of size for NPs beyond 2 nm. This key finding motivates the construction of thermodynamic tables for distribution coefficients using segregation observed with one or more NP sizes. The tables can enable accurate prediction of phase diagrams for nano materials across a wide-range of sizes. Key concepts of this new theory are demonstrated with Au-Pt, Ag-Au and Ni-Pt, which are found to exhibit complex size dependent segregation behavior for 2-6 nm NPs and relatively weaker size dependence beyond 6 nm. Numerically well converged values of Δ are calculated for small NPs using Monte Carlo (MC) simulations in the canonical ensemble. Simulations are based on an embedded atom method (EAM) potential for metal alloys.

Copyright © American Institute of Chemical Engineers. All rights reserved.

571928 Computing Solid-Liquid Interfacial Free-Energy Via Thermodynamic Integration

In this talk, we present a method to compute solid-solvent interfacial free-energies for sparingly soluble solutes via thermodynamic integration. The proposed method combines the phantom-wall method of Leroy and Muller-Plathe (J. Chem. Phys. 133, 044110 (2010)) and the Frenkel-Ladd method (J. Chem. Phys. 81, 3188 (1984))for solid-state free energy calculation. Our proposed method included the contribution of entropy of both solid and liquid phases in the computation of the interfacial tension. We apply this method to predict the influence of solvent on polymorph selectivity during crystal nucleation of molecular solids and vapor-liquid contact angles on a solid subtrate. The proposed method for computing interfacial tension can be easily implemented using the open-source molecular dynamics package LAMMPS. This can facilitate widespread adoption of this method by a large number of research groups.

Copyright © American Institute of Chemical Engineers. All rights reserved.

579709 Opening Comments for Session on Thermodynamics Needs of the Chemical Industry

A brief review of the history and objectives of AIChE's session on Thermodynamics Needs of the Chemical Industry.

Copyright © American Institute of Chemical Engineers. All rights reserved.

554065 Current State of Thermophysical Property Measurements and Reporting in Scientific Publications

Experimental thermophysical and thermochemical property data for chemical substances, mixtures, and materials are needed in various fields, such as chemical engineering, design of new processes and products, development of predictions/correlations, and education. Two factors are critical for addressing the above needs: availability and quality. Depending on their quality, the data can be useful or misleading.

While the amount of published thermophysical data constantly increases and their searchability improves, the reliability of the data in each case requires exploration. A set of methods and tools for data quality analysis is used at the NIST Thermodynamics Research Center (TRC) [1]. Application of those methods reveals the extent of inconsistencies and provides statistical information. For the journals cooperating with NIST/TRC for data validation, we apply our checks to accepted manuscripts and can compare their state before and after interaction with TRC. The aspects we consider are completeness of data reporting, physical/thermodynamic validity of data, and numerical representation of data. A significant part (up to 50%) of manuscripts has fixable problems such as incomplete reporting, ambiguities in identification of substances and properties, and typographical errors. Non-fixable problems such as flaws in technique, non-physical or inconsistent values are also frequently observed. The statistics of both will be provided in the presentation. While problems detected during the journal cooperation process are typically fixed after TRC provides a data report to the journals prior to publication, the rate of data problems in other journals is generally higher. Hence, no published property value can be blindly used without expert evaluation.

Part of TRC's activities are dedicated to improvement of the interaction with the cooperating journals. The experience gained in that process is used more broadly with the intent to increase the efficiency of thermophysical research and better serve the needs of data users. An element of those activities is the Good Reporting Practice initiative. Other planned actions will be dedicated to assisting reviewers and experimentalists with a goal to obtain more reliable results from different kinds of measurements.

This presentation will provide the statistics in more details, including an analysis of possible causes of the observed measurement and publication mistakes. It will also mention a potential effect of data errors on process simulation.

[1] Vladimir Diky, Ala Bazyleva, Eugene Paulechka, Joseph W. Magee, Vikina Martinez, Demian Riccardi, Kenneth Kroenlein. Validation of thermophysical data for scientific and engineering applications. J. Chem. Thermodyn, 2019, 133, 208-222.

*Corresponding author: *Email: ala.bazyleva@nist.gov;* Phone: +1-303-497-5981

Copyright © American Institute of Chemical Engineers. All rights reserved.

569240 An Improved Spectroscopic Method for Determination of Association/Solvation Parameters Used in Process Models

Process modeling of systems continues to be challenging when association and solvation are present. Accurate modeling of mixtures with such species will become increasingly important because renewable chemicals, fuel intermediates and biopolymers typically associate via hydrogen bonding. Equations of state such as SAFT, CPA and ESD have gained popularity by explicitly representing hydrogen bonding. In a previous study, we have developed an activity-coefficient thermodynamic model which incorporates Wertheim's perturbation theory (TPT-1). However, association parameters used in literature differ significantly depending on the type of data used to derive them. For example, the parameter for the self-association of methanol can vary by up to a factor of four.

This work focuses on the determination of the Wertheim association parameters using spectroscopic measurements such as FTIR. While this approach is not new, it is complicated by the challenges involved in quantifying the associating clusters such as alcohol monomers and dimers. Issues persist in curve fitting for evaluating the spectra where absorbances are highly concentration and temperature dependent. More importantly, the significant variation in the IR molar extinction coefficient across the broad hydroxyl band has been ignored in most studies and thus hindered accurate quantification of association. To address these challenges, we have developed a method based on quantum calculations for scaling experimental FTIR spectra such that more meaningful extraction of model-relevant thermodynamic constants is possible. The parameters acquired from our methods are compared to those obtained from other studies and used to model the phase equilibria of binary alcohol-alkane mixtures. The results demonstrate the capabilities of the activity coefficient model and further prove the importance of accurately capturing the hydrogen bonding phenomena in thermodynamic modeling.

Copyright © American Institute of Chemical Engineers. All rights reserved.

570182 A Molecular-Based Screening Tool for Water-Free Amine Solvents for CO_2 Capture

Currently, the chemical absorption of CO_2 using aqueous solutions of amines mainly, is the most mature technology for CO_2 capture and separation, that has been applied in many industrial processes. However, the high regeneration energy requirement is the major hurdle to the large-scale deployment of this technology. Among the various strategies present as a remedy for this issue, the utilization of water-lean or water-free solvents provide an attractive alternative. This is done through either partial (water-lean) or full (water-free) replacement of water with alternative physical solvents with lower heat capacity and evaporation enthalpy that, when combined with traditional alkanolamines, form physical-chemical absorbents. The accurate determination of CO_2 solubility in the utilized solvents is an integral factor in the design and simulation of carbon capture or acid gas treatment processes along with their techno-economic evaluation. However, the lack of experimental data on the thermophysical properties of new amine-based solvents or blends is a major obstacle for reliable process design and simulation. Therefore, devising a consistent and reliable thermodynamic model capable of accurately describing the phase equilibria, thermophysical and transport properties of CO_2 and amine-based systems is highly needed.

In this contribution, we demonstrate the capabilities of a robust molecular-based framework for the description of key thermophysical properties of water-free and water-lean amine-based solvents for the efficient removal of CO_2 and other acid gases from industrial gas streams at relevant gas separation process conditions. The screening tool is built on the molecular-based equation of state (EoS) soft-SAFT [1] combined with the Free-Volume Theory for the integrated modeling of phase behavior, enthalpies, densities and viscosities. Solvents investigated in this work include mixtures of various alkanolamines with different physical solvents such as alcohols, glymes, glycols, polar aprotic solvents and ionic liquids (ILs). Molecular parameters of all the investigated substances are either taken from previous contributions or developed in this work and used here in a transferable manner for describing the solubility of CO_2 in the selected solvents at conditions of relevance for industrial applications.

The description of the chemisorption of CO_2 in the studied amine + physical solvent blends is done in a consistent manner with that for the chemisorption of CO_2 in aqueous amine reported in our previous contributions [2,3]. A scheme of implicit reactions is used to describe the formation of carbamate and/or bicarbonate products resulting from the chemical reactions between CO_2 and the examined amines. The influence of the organic solvents on the chemical reactivity of amines with CO_2 in water-free solvents is investigated in detail. This procedure eliminates the need to specify the detailed equilibrium reactions and significantly reduces the number of parameters required to represent the absorption process. A maximum of two adjustable model parameters, optimized for a fixed amine concentration, suffice to represent the absorption of CO_2 in the studied solvents over a broad range of conditions, with high predictive and extrapolative capabilities. The acquired results from our the developed thermodynamic framework are compared to experimental data gathered from literature and predictions from the recommended thermodynamic models implemented in the Aspen Plus® (V.10) process simulator and the model found to be capable of providing reliable solubility estimations over a broad range of pressure, temperature and compositions.

The collection of these results demonstrates that once the robust molecular model is developed and validated, it can be used as a predictive tool, combined with macroscopic thermodynamics, as a reliable platform for the screening of chemical solvents as function of key process parameters, and as a valuable tool for process modelling and optimization.

This work is partially funded by the ADNOC Gas Processing and its stakeholders through the Gas Research Center (project GRC2018-003) and Khalifa University of Science and Technology (RC2-2019-007).

References

[1] F.J. Blas, L.F. Vega, *Mol. Phys*. 92, 135–150 (**1997**). F.J. Blas, L.F. Vega, *Ind. Eng. Chem. Res*. 37, 660–674 (**1998**).

[2] J.O. Lloret, L.F. Vega, F. Llovell, *J. CO_2 Util*. 21, 521–33 (**2017**).

[3] L.M.C. Pereira, F. Llovell, L.F. Vega, *Applied Energy*. 222, 687-703 (**2018**).

Copyright © American Institute of Chemical Engineers. All rights reserved.

571049 Relationship of Solubility and Thermodynamic Parameters for Amine-Containing Non-Regenerative H_2S Scavengers

Dan F. Smith Department of Chemical Engineering, Lamar University, P.O. Box 10053, Beaumont, Texas 77710, United States

Removal of H_2S from oil and gas streams has been a priority, oftentimes a necessity, for upstream crude oil/gas facilities for decades. However, the advances in production, namely horizontal drilling and hydraulic fracking, have produced oils and natural gases that contain higher concentrations of H_2S than previously seen. Crudes containing > 0.5 % sulfur are considered sour and must undergo separation and/or conversion processes to lower the H_2S content prior to entering the midstream pipelines and railcars. H_2S scavengers can be categorized into two types: regenerative and non-regenerative. Regenerative scavengers, such as alkanolamines, absorb H_2S using reversible absorption with stripping recovery processes. With non-regenerative scavengers, an irreversible reaction converts the H_2S to form sulfur-containing hydrocarbons. Non-regenerative scavengers are usually aldehydes, ethers, and amines (i.e. acrolein, glyoxal, formaldehyde, alpha-aminoethers, or triazines). Triazine-based scavengers have become the most widely used of the reactive-type scavengers. 1,3,5-hexahydrotriazine is the most frequently used; however, its byproducts equipment fouling, particularly overhead condensers within refineries.

These harmful effects have led to molecular adaptations to the basic triazine molecule whose byproducts would be benign to equipment fouling and corrosion. Varying side chains have the impact of making the triazine scavenger more (or less) soluble in aqueous (or non-polar) crudes. In addition, these side chains could be tailored to be more active for sour gas, rather than sour liquid crude streams. Side chain modifications lead to uncertainty of solubility, liquid/gas interactions, and possible byproduct complications. In this work, solubility and scavenger/ion interactions have been determined for triazines containing ethanolic, amino, bromine, and acrolein substituents, namely 1,3,5-Triacryloylhexahydro-1,3,5-triazine, cyanuric acid, and 1,3,5-Tris(2-hydroxyethyl)cyanuric acid. By adding salts to the mixture, the solubility data yields insightful information towards the ionic mediums in which these materials could be used. Moreover, Setchenow's coefficient, the octanol/water partition coefficients, and the activity coefficients have been determined to examine the potential fate these scavengers when considering environmental factors. The experimental results and model development will be presented.

Copyright © American Institute of Chemical Engineers. All rights reserved.

558709 Minimizing Process Development Costs in Industry By Computational Molecular Modeling Screening: Separation of Enantiomers for Pharmaceutical Applications

The overall goal of this project is to develop an efficient (time and cost) and reliable computational molecular modeling for screening conditions for effective separation of enantiomers using chiral stationery phases. More than half of all pharmaceuticals are chiral compounds. The potential market for enantiomer separation is $1.2 billion. Although enantiomers of chiral compounds have the same chemical structure, they can exhibit marked differences in physiological activity; therefore, it is important to remove the undesirable enantiomer. Chromatographic separation of chiral enantiomers is one of the best available methods to obtain enantio-pure substances, but the optimization of the experimental conditions can be very time-consuming. Generally, there are several chiral stationary phases to choose from; the most commonly used ones are polymers that have an amylose or cellulose backbone, with various sidechains, coated or covalently bonded to a solid support. Then there is the choice of the solvent system that constitutes the mobile phase. Furthermore, modifiers (acids or bases or buffers) may be added to the solvent in order to change or protect some functional groups on the drug to effect a better separation between the enantiomers. One could try different CSPs and mobile phases using educated guesses, but the number of possible combinations is huge, and could easily run into hundreds of combinations. This has stymied progress for developing new drugs especially orphan drugs. To overcome these problems, we have used molecular dynamics simulations to screen for these choices. Chromatographic enantiomer separation is a dynamic process, with the interactions between the drug and the chiral stationary phase mediated by the solvent, thus, no single interacting structure, such as could be found by finding low-energy structures via MD or quantum calculations, could possibly describe and account for the ratio of residence times in the chromatographic column for the enantiomeric pair. Atomistic molecular dynamics simulations permit us to develop an approach to predict which enantiomer elutes first and also provide an estimate of the separation factor.

Copyright © American Institute of Chemical Engineers. All rights reserved.

572458 Understanding Diffusion and Catalyst Deactivation in the Chemical Funneling of Lignin Derived Monomers over Zeolites

To design biorefineries capable of competing with established petroleum processes for producing organic chemicals, it is imperative that the use of all fractions of the biomass feed is optimized. Cellulose and hemicellulose have been investigated for use in many chemical and biological processes to generate bio-based organic molecules. However, lignin (the aromatic polymer occupying interstitial space between cellulose fiber bundles) has historically been underutilized as a low quality fuel rather than as a potential platform for aromatic organics. This attitude towards lignin has changed in recent years and new approaches are quickly being generated to try and make effective use of this resource in addition to cellulose and hemicellulose.

For these processes to utilize the entirety of the lignin stream in as cost effective a manner as possible, close attention must be paid to catalyst design and function. Understanding precisely how these catalysts behave on stream and what properties lead to longer catalyst activity will lead us to design better catalysts and reactor systems. The focus of this talk will be on exploring the effects of lignin-based aromatics, particularly oxygen containing species such as phenols, catechols, and guaiacols, on the kinetics of alkyl side chain cracking from lignin oligomers. We show that unusually fast deactivation is caused by strongly bound oxygenates, which hinder diffusion into and out of zeolite pores. We model how these road-blocking aromatics change diffusion within the zeolites and how this impacts conversion and catalyst lifetime. Finally, we study the influence of deactivation on the distribution of model lignin depolymerization products and propose process options, which allow for consistent activity and reliable conversion of such mixtures into easily separable products. This chemical funneling will be crucial for avoiding energy intensive separation in pursuit of drop-in petroleum replacement chemicals from biorefineries.

Copyright © American Institute of Chemical Engineers. All rights reserved.

568740 Tuning the Paraffin-to-Olefin Ratio in High-Octane Gasoline Synthesis Using Bimetallic BEA Zeolite Catalysts

Recent research on the production of HOG from DME over BEA zeolites suggests that this process offers an attractive alternative to traditional methanol-to-gasoline (MTG) technologies by operating at milder conditions and providing greater carbon efficiency[1]. The HOG product is rich in branched olefins and paraffins, particularly triptane (2,2,3-trimethylbutane) which has high research- and motor-octane numbers (RON 112, MON 101), making it an ideal unleaded fuel or fuel additive for aviation or racing applications[2]. RON and MON are the key fuel properties affecting engine anti-knock performance and are correlated with paraffin and olefin content[3].

On copper-modified BEA (Cu/BEA), ionic Cu(I) species are active for alkane dehydrogenation whereas metallic Cu species are responsible for hydrogenation chemistry[4]. Here we report on the effect of including ionic Zn or Ni species as a means of increasing the dehydrogenation activity of Cu/BEA, and thus enabling control over the paraffin to olefin ratio (P/O) in DME-to-HOG reactions. Bimetallic Cu-Zn/BEA and Cu-Ni/BEA catalysts were synthesized by first incorporating Ni or Zn by ion-exchange, followed by incipient-wetness impregnation of Cu. Similar metal loadings, acid site densities, and Bronsted/Lewis acid ratios were confirmed across all materials.

All catalysts were tested in DME-to-HOG reactions with co-fed H_2 at 200 °C, 15 psia, 1:1 DME:H DME flow of 2.2 g_{DME}-g_{cat}^{-1}-h^{-1}. Compared at similar turnover numbers (TON), Cu-Ni/BEA displayed similar activity to Cu/BEA however, it exhibited greater hydrogenation activity as evidenced by the higher C_{5-8} P/O ratio. Cu-Zn/BEA demonstrated decreased activity and the lowest P/O ratio, attributed to enhanced dehydrogenation activity at ionic Zn sites. For the HOG-range products (*i.e.,* C_{5-8} hydrocarbon species), each catalyst had an estimated RON of 98-99, while the MON decreased with decreasing P/O ratio. Compared to Cu/BEA, these bimetallic catalysts demonstrate a unique balance of hydrogenation and dehydrogenation of the hydrocarbon products to access markedly different P/O product ratios without requiring a separate unit operation or a mixed catalyst bed.

References

1. Tan, E.C., Talmadge, M., Dutta, A., Hensley, J., Snowden-Swan, L.J., Humbird, D., Schaidle, J., Biddy, M. "Conceptual process design and economics for the production of high-octane gasoline blendstock via indirect liquefaction of biomass through methanol/dimethyl ether intermediates" (2016) Biofuels, Bioproducts and Biorefining, 10 (1), pp. 17-35. DOI: 10.1002/bbb.1611
2. Khadzhiev, S.N., Magomedova, M.V., Peresypkina, E.G. "Triptane synthesis from methanol and dimethyl ether: A review" (2016) Petroleum Chemistry, 56 (3), pp. 181-196. DOI: 10.1134/S0965544116030063
3. Ghosh, P., Hickey, K.J., Jaffe, S.B. "Development of a detailed gasoline composition-based octane model" (2006) Industrial and Engineering Chemistry Research, 45 (1), pp. 337-345. DOI: 10.1021/ie050811h
4. Farberow, C.A., Cheah, S., Kim, S., Miller, J.T., Gallagher, J.R., Hensley, J.E., Schaidle, J.A., Ruddy, D.A. "Exploring Low-Temperature Dehydrogenation at Ionic Cu Sites in Beta Zeolite to Enable Alkane Recycle in Dimethyl Ether Homologation" (2017) ACS Catalysis, 7 (5), pp. 3662-3667. DOI: 10.1021/acscatal.6b03582

Copyright © American Institute of Chemical Engineers. All rights reserved.

573730 Evaluating Zeolite Supported Pt/ZIF-67 with Low Noble-Metal Loading for the Deoxygenation of Saturated and Unsaturated Fatty Acids

Biomass-derived transportation fuels provide the opportunity for both increased national energy security and reduced greenhouse gas emissions when compared to conventional petroleum fuels. Many processes exist for the production of fuels from biomass including pyrolysis, hydrothermal liquefaction, homogeneous catalysis, and heterogeneous catalysis. Our research is focused on the use of heterogeneous catalysts for the deoxygenation of abundant fatty acids found in vegetable oils, algae oils, and animal fats. The composition of many non-edible plants with high oil content is divided into saturated, unsaturated, and polyunsaturated fatty acids. Across the field of fatty acid upgrading, most researchers have focused on model compounds lauric (12:0), stearic (18:0), and oleic acid (18:1).

In the current study, we evaluate the activity of low platinum loading ZIF-67/zeolite 5A catalysts for the deoxygenation of polyunsaturated linoleic acid (18:2) under nitrogen, hydrogen, and carbon dioxide atmospheres. The results show that a higher degree of unsaturation can cause lower yields to desired alkanes due to the increased reductive demand of the polyunsaturated backbone. Overall, yields to heptadecane (C_{17}) from linoleic acid were as high as 65.3, 63.5, and 58.4% under H_2, CO_2, and N_2 atmospheres, respectively. Reactions in H_2 followed hydrodeoxygenation while N_2 and CO_2 resulted in decarboxylation pathways. Recyclability and textural properties of the catalyst were explored for all reactive atmospheres showing a slight decrease in surface area, good maintenance of crystallinity, and high stability at elevated temperatures by thermogravimetric analysis.

Copyright © American Institute of Chemical Engineers. All rights reserved.

560051 Phosphorus-Substituted Zirconium Metal-Organic Frameworks with Bronsted Acidity and Their Versatility for Biomass Catalysis

The conversion of biomass feedstocks to produce platform chemicals is an important alternative to petroleum-based resources. Dienes are high-volume chemicals used as precursors in the productions of polymers. They also find applications as surfactants and lubricants. Synthesizing dienes by dehydra-decyclization of saturated furans using Bronsted acid catalysts appear as a promising strategy. Phosphorus-modified materials have been reported to have Bronsted acidity with exciting applications for catalysis of biomass, including phosphorus-supported zeolites and oxides. MOF-derived solid acids have also been applied in catalysis and separations. In this work, zirconium metal-organic frameworks with the UiO-66 topology with high stability were synthesized. We report the introduction of phosphonate and phosphonic acid moieties to both 1,4-benzenedicarboxylate (BDC) linker and biphenyl-4,4'-dicarboxylate (BPDC) synthesized by organic chemistry protocols to form UiO-66-PO_3H_2 and UiO-67-PO_3H_2 by a post-synthesis modification method. The samples had their physicochemical properties characterized by different methods. While the P-MOFs retained crystallinity, porosity, morphology, and stability of the parent materials, phosphorus was successfully incorporated into the framework. Bronsted acidity was characterized by isopropanol dehydration and Hoffman elimination of amines. Catalytic activity was probed in the gas phase dehydra-decyclization of furans. The synthesized MOFs showed activity and selectivity to dienes, indicating their potential application as heterogeneous acid catalysts.

Copyright © American Institute of Chemical Engineers. All rights reserved.

570421 Upgrading Glycerol Via Selective Hydrodeoxygenation Reaction on Copper-Modified Molybdenum Carbide and Nitride Surfaces

Biodiesel production is a promising approach to address the current environmental issues and reduce our reliance on fossil fuels. During biodiesel production, a significant amount of glycerol is produced. As the production of biodiesel increases, a large surplus of glycerol is generated. Therefore, the upgrading of glycerol can largely benefit the process economy. Among the various pathways for glycerol upgrading, the selective hydrodeoxygentation (HDO) reaction is particularly interesting, in that different numbers of C-O bond of glycerol can be cleaved to form value-added products.[1] In this work, the active sites on the Cu/Mo_2C surface for selectively cleaving certain numbers of C-O bond of glycerol were revealed, demonstrating the feasibility of tuning product selectivity by simply changing the Cu coverage.[2] Since carbon tends to accumulate on the Mo_2C surface during the carburization process, molybdenum nitride (Mo_2N) was identified as an alternative for Mo_2C. The extra surface nitrogen species during the reaction of Mo with NH_3 can desorb as N_2, leaving a clean surface for better interaction between Cu modifier and Mo. Compared to Mo_2C and Cu/Mo_2C, Mo_2N and Cu/Mo_2N show similar activities with added functionalities. This work develops a fundamental understanding of Cu-modified carbide and nitride surfaces, which should in turn provide insights into the rational design of efficient catalysts for the selective HDO reaction of biomass-derived oxygenates.

References

1. Z. Lin, R. Chen, Z. Qu and J. G. Chen, Green Chem., 2018, 20, 2679–2696.
2. W. Wan, S. C. Ammal, Z. Lin, K.-E. You, A. Heyden and J. G. Chen, Nat. Commun., 2018, 9, 4612.

Copyright © American Institute of Chemical Engineers. All rights reserved.

573611 Investigating the Role of Oxophilic Modifiers in Bimetallic Catalysts for Enhanced Hydrodeoxygenation

Introduction

In the search for renewable energy sources, lignocellulosic biomass has emerged as a promising alternative to typical petroleum-based fuels. However, due to the complex and recalcitrant nature of the lignin fraction, current catalyst technology is inefficient for upgrading bio-oil to liquid fuels and value-added products. Recent advances have demonstrated that bimetallic catalysts consisting of a noble metal and an oxophilic modifier may improve catalyst performance while resisting deactivation under the harsh conditions of bio-oil upgrading[1,2]. Several theories have been proposed for the role of the modifier species in the hydrodeoxygenation (HDO) of lignin-derived oxygenates. Some groups have suggested that the increased performance derives primarily from an electronic interaction between the oxophilic metal and the noble metal[3]. Others have hypothesized that each metal species performs a unique function in the HDO reaction. In this case, the modifier species may act as an anchoring site for the oxygen-containing functionality, with the strength of the metal-oxygen interaction being a key predictor for catalyst performance[4]. Alternatively, the modifier may act primarily as a Bronsted or Lewis acid site, in which case the strength and density of acid sites on the surface would be expected to correlate with improved HDO activity[5]. Thorough characterization of the reactivity and surface structure of multiple bimetallic catalysts is needed to pinpoint which factors can be used most reliably to predict their performance in HDO upgrading.

Materials and Methods

In order to study the effect of a modifier on the HDO activity of bimetallic catalysts, samples of Pt-W catalysts were synthesized at varying modifier loadings through incipient wetness impregnation of a γ-Al_2O_3 support. Reaction experiments were carried out in a tubular plug-flow reactor using benzyl alcohol as the probe molecule. Surface characterization was examined through inductively coupled plasma – mass spectrometry (ICP-MS), CO chemisorption, CO diffuse reflectance infrared Fourier-transform spectroscopy (DRIFTS), temperature programmed reduction (TPR), and pyridine DRIFTS.

Results and Discussion

Recent results demonstrate that bimetallic Pt-W/Al_2O_3 samples show improved HDO activity with increasing W loading up to about 6 wt.% W. However, samples with higher W loadings experience a sharp reduction in overall catalytic activity, as seen in Figure 1. Additionally, the selectivity towards the undesirable decarbonylation (DC) product, benzene, is suppressed at these high W loadings. This behavior suggests that the presence of W in these bimetallic systems introduces at least two competing factors that affect the HDO activity of the catalyst.

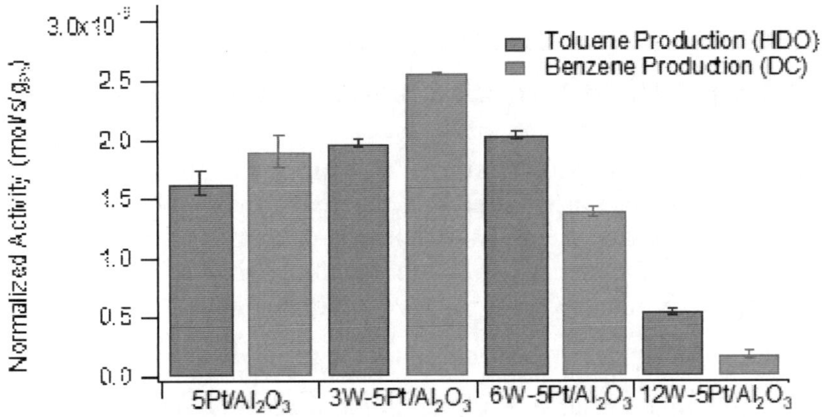

Figure 1. Benzyl alcohol HDO and DC activity for bimetallic Pt-W/Al_2O_3 samples

Further characterization of the physicochemical properties of the surface have offered hints as to which factors are correlated with increases and decreases in catalyst performance. For example, pyridine DRIFTS has revealed that Brønsted acid site density is strongly correlated with HDO activity for our Pt-W/Al_2O_3 catalysts while Lewis acid site density has no correlation. Additionally, CO DRIFTS of these samples indicate that W species are in close coordination with Pt at all weight loadings. In conjunction, these results imply that the W in our samples tends to coordinate with Pt nanoparticles on the surface of the catalyst, which may result in the formation of new metal-acid bifunctional sites that are active for the HDO of benzyl alcohol. However, despite these beneficial interactions at moderate W loadings, the results of CO chemisorption suggest that high W loadings tend to reduce the catalytically active surface area by covering Pt atoms.

Significance

HDO of aromatic oxygenates is a necessary step in the upgrading of biomass to alkane fuels and products, and bimetallic catalysts have shown promise for activating this process while minimizing side products. However, the fundamental role of each metal species in this reaction is currently poorly understood. The identification of key factors for improving catalyst performance would allow for the rational design of bimetallic catalysts with optimized activity and selectivity for desired products.

References

1. Robinson, A. M., Hensley, J. E. & Medlin, J. W. *ACS Catal.* **6,** 5026–5043 (2016).
2. Alonso, D. M., Wettstein, S. G. & Dumesic, J. A. *Chem. Soc. Rev. Chem. Soc. Rev* **41,** 8075–8098 (2012).
3. Sun, J. *et al. J. Catal.* **306,** 47–57 (2013).
4. Robinson, A. M., Mark, L., Rasmussen, M. J., Hensley, J. E. & Medlin, J. W. *J. Phys. Chem. C* **120,** 26824–26833 (2016).
5. De Souza, P. M. *et al.* Hydrodeoxygenation of Phenol over Pd Catalysts. *ACS Catal.* **7,** 2058–2073 (2017).

Copyright © American Institute of Chemical Engineers. All rights reserved.

575106 Enhanced Aromatic Selectivity during Deoxygenation of Phenolic Model Compounds over Bifunctional Catalysts

Phenolic compounds are produced either from fast pyrolysis of lignocellulosic biomass to produce biooils or from depolymerization of woody or grass lignins. To efficiently convert lignin to meet the specifications of either transportation fuels or aromatic chemical precursors, catalytic upgrading via hydrodeoxygenation (HDO) is a necessary step. Among the diversity of phenolic compounds present in either biooil or depolymerized lignin, guaiacol is often used as a model compound, since it includes both phenolic (Ph -OH) and methoxy (-OCH) groups, representing a large number of oxygen containing components in these sources. In the present work, various bifunctional catalysts with metal and acidic sites were synthesized and evaluated for guaiacol HDO performance with the target of maximizing aromatic hydrocarbon selectivity. It was found that under optimum operating conditions, guaiacol is fully deoxygenated with stable conversion and selectivity profiles over bifunctional catalysts, leading to a maximum aromatic hydrocarbon yield of ~35%. Full deoxygenation of phenolic compounds to aromatic hydrocarbons with enhanced selectivity is demonstrated with bimetallic catalysts supported on mesoporous silicates with tunable acidity. The results have both fundamental and practical implications for designing catalysts and processes for selective deoxygenation of phenolic compounds in either biooil or depolymerized lignin mixtures.

Copyright © American Institute of Chemical Engineers. All rights reserved.

577982 Kinetic Hydrate Inhibitors: Gas Hydrate Induction Time Probability Distributions and Deviations from Classical Nucleation Theory

As subsea oil and gas production moves toward deeper water, high-pressure and low-temperature conditions leads to increased risk of gas hydrate formation. The agglomeration and aggregation of hydrate particles can then lead to costly pipeline plugging events. Various inhibitor chemicals can be injected into flowlines to prevent or delay hydrate formation. Unlike Thermodynamic Hydrate Inhibitors (THIs) such as methanol however, Kinetic Hydrate Inhibitors (KHIs) do not affect the hydrate phase equilibrium curve, rather they delay hydrate formation. Robust KHI performance qualification thus requires formation itself to be measured rather than the dissociation curve (which can be used to qualify THI performance). This poses a challenge since hydrate nucleation is stochastic and thus necessitates a high number of independent measurements of hydrate formation events, a requirement which is challenging to realise with conventional apparatus that are only capable of relatviely slow temperature ramping.

Here we detail the use of a high-pressure-stirred-automated-lag-time-apparatus (HPS-ALTA) to enable efficient quantification of methane hydrate formation probability in the presence of a KHI. Typically 100 independent hydrate formation events are obtained for each experimental condition, enabling the construction of smooth induction time probability distributions. We confirm previous findings in that we demonstrate that increasing KHI concentration leads to increased induction times, albeit with diminishing returns as the concentration is increased in the (1-2) wt% range. We also show that although induction times distributions are exponentially distributed for the KHI-free system (consistent with predictions from Classical Nucleation Theory), this is not the case in the presence of a KHI where our experimental data can be best described by a gamma distribution with lower relative standard deviation. This result has significance not only for understanding hydrate formation risk in the presence of KHIs but also for improving our theoretical understanding of the mechanism by which KHIs delay hydrate formation.

Copyright © American Institute of Chemical Engineers. All rights reserved.

571307 Structure H Hydrate Formation with Lmgs and CO_2, in Ice and Water

Structure H hydrate formation with LMGS and CO_2, in Ice and Water

Namrata Gaikwad[1], Omkar S. Kushwaha[1], Praveen Linga[3], Jitendra Sangwai[2], Rajnish Kumar[1]*

[1] Department of Chemical Engineering, Indian Institute of Technology, Chennai, 600036, India

[2] Department of Ocean Engineering, Petroleum Program, Indian Institute of Technology, Chennai, 600036, India

[3] Department of Chemical and Biomolecular Engineering, National University of Singapore, 117585, Singapore

Corresponding author: rajnish@iitm.ac.in

Keywords: Gas hydrates, CO_2 Capture, CH_4 Storage, Hydrate Kinetics, Hydrate Dissociation.

Abstract

Van der Waals radius of carbon dioxide molecules are significantly larger than methane and thus it is a reluctant help gas for sH hydrate formation. A successful attempt has been made to form structure H (sH) hydrate at low pressure from four different types of experiments in ice and water by employing methane and carbon-dioxide as a small molecule guest substance (SMGS) and polar tert butyl methyl ether (TBME) and hydrophobic cyclo-octane as a large molecule guest substance (LMGS). One of the ideas was to use different LMGS such that the effect of oxygen molecule containing LMGS which are slightly polar could be studied with those LMGS which are completely non-polar. Type-1 experiments were carried out to study the kinetics of sH hydrate formation and dissociation for methane as SMGS for two identified sH formers at 274K. Type-2 experiments were performed to study the kinetics of carbon dioxide sH hydrate formation and dissociation at 274K i.e, sH hydrate formation with liquid water, such data are very rare in literature! Type-3 experiments were studied to understand the formation and dissociation kinetics for carbon dioxide sH hydrate at 266K, the literature suggests that sH hydrate formation with CO_2 and LMGS is possible only at sub-zero temperature, i.e., with solid water. Type-4 experiments were done to study the effect of driving force on carbon dioxide sH hydrates formation in ice at 266K. Experimental data on avg. gas uptake, water to hydrate conversion, kinetics of methane and carbon dioxide sH hydrate formation and dissociation at low pressure has been obtained. This study shows that kinetics of gas uptake and water to hydrate conversion depends on the type of LMGS and driving force used. Experiments with

TBME as a LMGS have shown higher gas uptake and rapid hydrate formation for methane. However, the rate of hydrate growth for CO_2 hydrate was faster with TBME in ice and cyclo-octane in water. It is observed that CO_2 molecules interact strongly with TBME in the liquid phase which might be acting as an inhibitor in the formation of hydrate. The system containing TBME as a LMGS has displayed a higher dissociation rate compared to cyclo-octane for both methane and carbon-dioxide gas.

Copyright © American Institute of Chemical Engineers. All rights reserved.

575481 Understanding the Influence of Thermodynamic Hydrate Inhibitors on the Mechanics of Cyclopentane Hydrate Films Using Interfacial Rheology with Subphase Exchange

Clathrate or gas hydrates are known to form at water-hydrocarbon and water-natural gas interfaces, generally at low temperatures and high pressures, to trap the guest hydrocarbon or gas molecules in crystalline cages of ice-like water. Gas hydrates are of interest to the oil and gas recovery community as rapid nucleation and growth of these solids can block flow lines and create significant problems. Thus, extensive efforts have been made to link hydrate film nucleation, growth, and dissociation using bulk hydrate formation and measurement techniques. However, the link between film nucleation, growth, and dissociation remains unclear because of the difficulties associated with generalizing bulk results to processes comprised of interfacial phenomena. Recently, interfacial rheological methods have been employed to investigate the growth kinetics and film mechanics of cyclopentane hydrate films that conveniently form at atmospheric pressure. By modifying our interfacial rheology setup to include the ability to exchange subphase fluid without disturbing the interface, we are able to measure hydrate film mechanical properties before, during, and after the introduction of thermodynamic hydrate inhibitors to gain fundamental insight into the interfacial processes associated with film nucleation, growth, and dissociation. We will present our results on the formation and dissociation of cyclopentane hydrate films in the presence of different thermodynamic inhibitors and discuss how this might influence bulk rheological properties.

Copyright © American Institute of Chemical Engineers. All rights reserved.

570674 Insight into the Hydrate Nucleation of Structure II Propane Hydrates from Molecular Simulation

Among the three main types of gas hydrate crystal structure, structure II (sII) hydrates are the most prevalent in oil and gas pipe line as propane and butane are commonly present, and these these molecules can only fit into the large cage of sII. While the nucleation of structure I hydrates has been been studied by a number of groups, there is still no insight into how sII hydrates forms, in particular when only a large guest, such as propane, is present to fill the large cages while leaving all the small cages empty. We performed molecular dynamic simulations for the nucleation of propane hydrates to capture the incipient cages of sII hydrates and to investigate the nucleation mechanism and crystallization process of sII hydrates. Results will be shown detailing the cages formed during the nucleation and the ordering leading to the crystalline structure.

Copyright © American Institute of Chemical Engineers. All rights reserved.

559776 Modeling the Effects of Interfaces on the Nucleation of Gas Hydrates and Ice in Porous Media

Gas hydrates are water crystalline structures stabilized by the presence of gas molecules. Gas hydrates occur naturally in the permafrost or in ocean-floor sediments at low temperatures and high pressures. Predictions on the amount of methane trapped within gas hydrates contained in marine sediments amounts to three orders of magnitude the worldwide conventional natural-gas reserves. Gas hydrates are an untapped energy resource and a potential contributor to global warming. Mathematical modeling of the stability, formation and dissociation of gas hydrates within sediments is a necessary tool for their management. Due to their similarity to ice, aspects of the behavior of gas hydrates can be modeled similarly to the treatment of ice.

Heterogeneous nucleation takes place during the formation of gas-hydrates and ice in porous media. The presence of surfaces and interfaces within porous media can effectively lower the energy barrier for nucleation. The present work will introduce mathematical models developed in order to describe the effects on energy barriers during heterogeneous nucleation of ice or gas hydrates in porous media. Three different types of interfaces found in these environments will be discussed: a) gas/liquid interfaces, b) solid surfaces, and c) gas/liquid/solid triple boundary lines.

Model predictions suggest that nucleation is more energetically favorable on solid surfaces than at the gas/liquid interface or triple boundary line. In the case of solid surfaces, nucleation is favored within concave surfaces (e.g., inside porous spaces) and to a lesser extent on convex surfaces (e.g., particulates), while flat surfaces act as the behavioral boundary between these two cases. Additionally, hydrophilicity of solid surfaces enhances the positive effect of these surfaces towards nucleation. In the case of gas/liquid interfaces and triple boundary lines, nucleation and crystal growth occur preferentially within the liquid phase rather on the gas phase, and it is favored when the gas/liquid interfaces are present between slightly hydrophobic surfaces.

Copyright © American Institute of Chemical Engineers. All rights reserved.

570727 A Fundamental Comparison of the Stability and Dissociation Aspects of Structure I and Structure II Hydrates from the Perspective of Feasible Natural Gas Storage

The hydrate based approach to storing natural gas (known as Solidified Natural Gas (SNG) technology) has been gaining popularity in recent times owing to its environmentally benign operation, large volumetric capacity and non-explosive character. By transitioning from sI to sII hydrates through the introduction of thermodynamic promoters like tetrahydrofuran (THF) to the system, researchers have not only been able to favourably tune the thermodynamic operating conditions for hydrate formation (significantly higher temperature and lower pressure for hydrate formation), but also demonstrated remarkable kinetic performance during hydrate formation at the revised operating conditions. However, the true feasibility of this approach lies in the long term stability and thus storability of mixed CH_4-THF sII hydrates at atmospheric pressure and how it compares with that for pure CH_4 sI hydrate under similar storage conditions.

According to thermodynamic data, the equilibrium temperature for mixed CH_4-THF hydrates (considering 6 mol % of THF in the system) at 0.12 MPa is 277.7K, while the same for pure CH_4 hydrate at 0.1 MPa is 193 K. Thus ideally, for storage at atmospheric pressure and moderately low temperatures (268-271 K), mixed CH_4-THF sII hydrate should be able to demonstrate complete stability while the pure CH_4 sI counterpart should dissociate rather rapidly. In the current study - a first of its kind, a like for like comparison was made for the stability and dissociation characteristics of methodologically identical pure CH_4 (sI) and mixed CH_4-THF (sII) hydrates. Cylindrical sI and sII hydrate pellets were synthesized using a one of a kind SNG technology prototype. Synthesized hydrate pellets were transferred to storage units for seven to ten day stability tests at atmospheric pressure and moderately low temperatures (~268-271 K). Rate of hydrate dissociation via thermal stimulation using room temperature water stream was separately studied for the synthesised sI and sII pellets. Results indicate that while sII pellets were completely stable at the investigated storage conditions, sI pellets started dissociating as soon as the tests were initiated. This demonstrates clearly the remarkable stability of sII hydrates when stored at atmospheric pressure and moderately low temperatures, thus highlighting a major downstream advantage of opting for the mixed sII hydrate route as far as SNG technology is concerned. The results obtained from this study are of great importance from both technical and economic standpoints and should go a long way in ensuring SNG technology as an attractive alternative for mid to long term storage of natural gas.

Copyright © American Institute of Chemical Engineers. All rights reserved.

570540 Container-Free Gas Hydrate Nucleation Using High-Pressure Acoustic Levitation

Understanding hydrate nucleation is of direct relevance to the field of flow assurance in the oil and gas industry. However, due to its stochastic nature and potential apparatus dependencies, hydrate nucleation can be difficult to generalise. Conventionally, studies are performed with high-pressure vessels wherein solid-liquid interfaces generate potential hydrate nucleation sites that are non-trivial to characterise (e.g. microscopic imperfections). To overcome such constraints, we demonstrate here the first statistically significant set of truly container-free hydrate formation measurements using a high-pressure acoustic levitator. In our experiments, water droplets with well-defined geometries were acoustically levitated within a natural gas atmosphere at approximately 65 bar and 7 °C. Hydrate formation on the interface of the ellipsoidal droplets was directly imaged and used to extract induction times. Cumulative probability distribution functions for these induction times at a controlled sub-cooling were constructed and compared with those obtained from conventional measurements and with predictions from classical nucleation theory. Hydrate dissociation within the acoustic field was also compared to that measured in a high-pressure autoclave.

Copyright © American Institute of Chemical Engineers. All rights reserved.

572499 Molecular Design of Polymer Nanocomposites

Polymer nanocomposites (PNCs) are an exciting class of hybrid materials that contain typically inorganic nanoparticles embedded in a polymer matrix, and this class of materials has attracted a lot of interest due to their promise in numerous technologies. As one example, in some membrane applications polymer nanocomposites have been shown to be capable of breaking the typical tradeoff between selectivity and permeability that is observed, and there is wide interest in using PNCs as the matrix in structural materials. However, despite their promise, there are numerous fundamental challenges that remain in the design of polymer nanocomposites. The thermodynamics of even simple polymer nanocomposites remain poorly understood, and direct mappings between theoretical and experimental phase diagrams are rare in the field. Furthermore, strategies for reaching high nanoparticle loadings are few. In this talk, I will describe recent efforts to address both the need to predict the thermodynamics of polymer nanocomposites and the properties of nanoparticles at high loadings. Our group has both developed a novel suite of field-theoretic simulations techniques to study inhomogeneous polymer/nanoparticle composites, which enable the prediction of phase diagrams for this class of hybrid materials, and we have recently extended our methods to capture beyond-equilibrium phenomena to describe systems where processing plays a key role in the resulting structure. Separately, we have characterized the properties of a novel class of nanocomposites where the nanoparticles are loaded at high concentrations, up to random close packing. In these materials, the entropy of the polymer in this highly confined geometry plays a key role in both the formation and the properties of the composite. Finally, I'll conclude by highlighting outstanding challenges in the field that need to be addressed for both simulation and experiment.

Copyright © American Institute of Chemical Engineers. All rights reserved.

569491 Coarse-Grained Models for Predicting Structure and Thermodynamics in Macromolecular Systems with Specific and Directional Intermolecular Interactions

Recent work in my research group has been aimed at developing predictive coarse-grained (CG) models for investigating structure and dynamics in soft materials with specific and directional molecular interactions. Although computational studies have been tremendously useful in understanding molecular phenomena and guiding synthesis and engineering of new polymers for a wide variety of applications, the inability to capture small scale specific and directional interactions (e.g., hydrogen bonds) alongside macromolecular length and time scales represents a key limitation of most polymer simulation studies to date. We address this limitation by developing coarse-grained models that capture the anisotropic, directional and specific interactions (e.g., hydrogen bonding interaction) governing the structure and thermodynamics in many polymer systems of interest. We have been using molecular dynamics simulations with these new coarse-grained models for studies of biomaterials and polymer nanocomposites. In this talk, I will discuss a few examples of how the development of these CG models is enabled by synergistic feedback from concurrent/past experiments which I will highlight alongside the computational results from my group.

Copyright © American Institute of Chemical Engineers. All rights reserved.

562997 Computational Design of Water-Mediated Interactions

Hydrophobic interactions are a fundamental driving force for self-organization and self-assembly in biology, and key players in the production and performance of many synthetic materials. From a theoretical point of view, both hydrophobic solvation and interactions have been studied extensively for model solutes and interfaces. However, many molecules of interest are highly heterogeneous in both chemistry (e.g., with both hydrophobic and hydrophilic groups) and conformations (e.g., through configurational fluctuations). Such heterogeneities produce a rich potential design space that can be leveraged to create surfaces and solutes with optimized thermophysical behaviors. In this talk, I discuss our recent work to understand and to design heterogeneities that impact water, solute partitioning and solubility, and complex water-mediated interactions.

Copyright © American Institute of Chemical Engineers. All rights reserved.

563013 Characterizing Protein Hydration to Inform Its Interactions

The extent to which the inherent structure of water is perturbed by complex molecules, such as proteins, peptides, and surfactants, influences the thermodynamics and the kinetics of their assembly. However, accurately characterizing this perturbation is challenging, because the manner in which proteins disrupt the inherent structure of water depends not only on the chemistry of the underlying protein surface, but also on the precise topographical and chemical pattern displayed by the protein. Nevertheless, understanding the role of water in protein interactions is essential to understanding, predicting, and eventually controlling such interactions, which play a crucial role in the development of therapeutic strategies and in protein separations. In this presentation, I will discuss our recent successes in quantitatively characterizing the disruption of water structure in the hydration shell of proteins, and in using this information to predict the interfaces through which proteins interact with one another and self-assemble. Our approach also informs strategies for optimally modulating protein interactions, and facilitates the design of ligands that will bind to proteins of interest with high affinity and specificity. We hope that these advances will pave the way for the discovery of novel therapeutics that specifically target proteins of interest, and the rational design chromatographic ligands for challenging protein separations.

Copyright © American Institute of Chemical Engineers. All rights reserved.

560257 Deep Learning of Slow Collective Variables to Understand and Accelerate Biomolecular Folding and Assembly

The variational approach to conformational dynamics provides a framework to estimate from time series data the leading eigenfunctions of the transfer operator that propagates the system dynamics. In the context of molecular simulation, this allows the estimation of the slowest collective modes governing the microscopic time evolution of the system from molecular dynamics trajectories. These collective variables are extremely valuable in understanding the important facets of protein folding and self-assembly, and in furnishing good coordinates in which to perform accelerated sampling. We introduce a novel neural network architecture and loss function that we term state-free reversible VAMPnets (SRVs) to estimate within a variational approximation the full orthogonal spectral hierarchy of transfer operator eigenfunctions. Applications to the Trp-cage and WW domain mini-proteins and the hybridization of short DNA-oligomers identifies the slow pathways governing molecular folding and provides an optimal basis set for the construction of Markov state models (MSMs) with higher temporal resolution and state decomposition than state-of-the-art techniques. This approach unveils new understanding of the metastable states and folding and assembly pathways of these systems.

Copyright © American Institute of Chemical Engineers. All rights reserved.

572472 A Deep Neural Network Model for MD-Level Packing Density Predictions and Its Application in the Study of 1.5 Million Organic Molecules

The process of developing new compounds and materials is increasingly driven by computational modeling and simulation, which allow us to characterize candidates before pursuing them in the laboratory. One of the non-trivial properties of interest for organic materials is their packing in the bulk, which is highly dependent on their molecular structure. By controlling the latter, we can realize materials with a desired density (as well as other target properties). Molecular dynamics simulations are a popular and reasonably accurate way to compute the bulk density of molecules, however, since these calculations are computationally intensive, they are not a practically viable option for high-throughput screening studies that assess material candidates at a massive scale. In this work, we employ machine learning to develop a data-derived prediction model that is an alternative to physics-based simulations, and we utilize it for the hyperscreening of 1.5 million small organic molecules as well as to gain insights into the relationship between structural makeup and packing density. We also use this study to analyze the learning rate of the employed neural network approach and gain empirical data on the dependence of model performance and training data (size), which will inform future investigations.

Copyright © American Institute of Chemical Engineers. All rights reserved.

574614 Toward Better Solar Cells: First Principles Understanding of the Water Instability of Hybrid Perovskites

Organometal halide perovskites (OMHPs), composed of organic molecular cations encaged in inorganic framework (anions), are a center of attention in the renewable energy community as photovoltaic (PV) materials, LEDs, and high-gain photodetectors. Low-cost and low-temperature production methods, high optical adsorption over a broad solar spectrum, tunable bandgap, and long charge carrier lifetime and diffusion length have made this class of materials exceptionally attractive. Since 2009, their power conversion efficiency (PCE) has grown from 3.8%, to above 23%. Methylammonium lead iodide ($MAPbI_3$) is one OMHP that has been widely studied. Moisture can significantly catalyze and accelerate the degradation of $MAPbI_3$ creating a prominent obstacle to the commercialization of $MAPbI_3$. Here, using first-principles density functional theory (DFT) calculations we reveal how water incorporation in $MAPbI_3$, catalyzes the phase transition to the ($MAPbI_3.H_2O$ edge-sharing) monohydrate (colorless) phase, eliminating its favorable photovoltaic properties [1]. First, using extensive electronic structure analyses the fundamental chemical and electrostatic interactions between water and each component of $MAPbI_3$ (organic MA molecule and Inorganic PbI_3 octahedron) are investigated, demonstrating the nature of the bonding and its dependence on water concentration. Second, the energetics and kinetics of incorporated water is explored, leading to the discovery of spontaneous phase segregation into dry regions and regions with more than one water per formula unit—termed the "super-hydrous state." Third, the properties of the super-hydrous state including the acceleration of octahedron breaking and rearrangement by the high water density are analyzed. We elucidate for the first time that the degradation is a bulk rather than surface process, initiated at the super-hydrous regions. We discuss how this bulk super-hydrous degradation model can explain disparate recent experimental observations concerning the water-induced transition from (black) perovskite to edge-sharing PbI_2 (yellow) phase, and allows one to design schemes to alleviate the water instability of not only $MAPbI_3$, but the bigger class of hybrid organic-inorganic perovskites, paving the way for their commercialization as very efficient solar cells.

In addition to the practical applications that such first principles understanding can have for this class of perovskite solar-cells; our findings can have broader impact: the fundamental knowledge gained on the nature of water (lone-pair) interaction with other molecular groups or surface/bulk binding sites [1,2] can be generally used for understanding and prediction of water--materials interactions, wetting or materials' moisture stability.

[1] Arvin Kakekhani, Radhika N. Katti, Andrew M. Rappe "Water in hybrid perovskites: Bulk $MAPbI_3$ degradation via super-hydrous state." APL Materials 7, 041112 (2019)

[2] Arvin Kakekhani, Luke T. Roling, Ambarish Kulkarni, Allegra A. Latimer, Hadi Abroshan, Julia Schumann, Hassan AlJama, Samira Siahrostami, Sohrab Ismail-Beigi, Frank Abild-Pedersen, Jens K. Nørskov. "Nature of Lone-Pair–Surface Bonds and Their Scaling Relations." Inorganic chemistry 57, no. 12 (2018): 7222-7238.

Copyright © American Institute of Chemical Engineers. All rights reserved.

572474 Simulation of the Surface Tension of a Nematic Liquid Crystal in Contact with a Planar Solid

Simulation of the surface tension of a nematic liquid crystal in contact with a planar solid

P. E. Brumby[1], H. H. Wensink[2], A. J. Haslam[3], and George Jackson[3]

[1]Department of Mechanical Engineering, Keio University, 3-14-1 Hiyoshi, Kohoku-ku, Yokohama, 223-8522, Japan

[2]Laboratoire de Physique des Solides, Université Paris Sud & CNRS, 91405 Orsay Cedex, France

[3]Department of Chemical Engineering, Imperial College London, South Kensington Campus, London, SW7 2AZ, United Kingdom

Corresponding author e-mail: g.jackson@imperial.ac.uk

Knowledge of the surface tension between a liquid crystalline phase and solid substrate is of prime importance in broad areas of the physical and biological sciences ranging from gaining an understanding of the organisation of molecules in optoelectronic devices to the microstructures formed in living organisms. The direct molecular simulation of the interfacial tension of high-density liquid crystals in contact with solids is particularly challenging. The usual methodology involves the use of a thermodynamic route based on integrating the Gibbs adsorption equation where the change in the interfacial tension is related to the change in the chemical potential for a given value of the adsorption. This type of thermodynamic approach requires knowledge of the chemical potential with simulations in the grand canonical ensemble or from free-energy integration which is computationally prohibitive for high-density ordered states. In our current work we employ an efficient and versatile free-energy perturbation method based on the calculation of the components of the pressure tensor that it is applicable to non-spherical and non-convex molecules from a series of 'ghost' anisotropic test-volume deformations [1]. For inhomogeneous systems this method provides a particularly convenient route to the calculation of the interfacial tension (surface free energy) from molecular simulations [2]. We study the structural properties and interfacial tension of a fluid of hard-spherocylinder rod-like particles in contact with hard structureless at walls by means of Monte Carlo simulation. The calculated surface tension between the rod fluid and the substrate is characterized by a non-monotonic trend as a function of bulk concentration (density) over the range of isotropic bulk concentrations. A surface-ordering scenario is confirmed from our simulations: the local orientational order close to the wall changes from uniaxial to biaxial nematic when the bulk concentration reaches about 85% of the value at the onset of the isotropic-nematic phase transition. The surface ordering coincides with a wetting transition whereby the hard wall is wetted by a nematic film. Accurate values of the fluid-solid surface tension,the adsorption, and the average particle-wall contact distance are reported (over a broad range of densities into the dense nematic region for the first time), which may serve as a useful benchmark for future theoretical and experimental studies on confined rod fluids. The simulation data are supplemented with predictions from

[1] P. E. Brumby, A. J. Haslam, E. de Miguel, and G Jackson, "Subtleties in the calculation of the pressure and pressure tensor of anisotropic particles from volume-perturbation methods and the apparent asymmetry of the compressive and expansive contributions," *Molecular Physics*, **109**, 169-189 (2014)

[2] P. E. Brumby, H. H. Wensink, A. J. Haslam, and G. Jackson, "Structure and interfacial tension of a hard-rod fluid in planar confinement," *Langmuir*, **33**, 11754-11770 (2017).

Copyright © American Institute of Chemical Engineers. All rights reserved.

578120 Multi-Scale Simulation of Ionic Polyimide Composite Membranes Based on Structural and Electrostatic Maps

Polyimides are at the forefront of advanced membrane materials for CO_2 capture and gas purification processes. Recently, ionic polyimides (i-PIs) have been reported as a new class of condensation polymers which combine structural components of both ionic liquids (ILs) and polyimides through covalent linkages. In this work, the CO_2 separation characteristics of ionic polyimides are modeled using a multi-scale simulation strategy by combining density-functional theory (DFT), molecular dynamics (MD) simulations, grand canonical Monte Carlo (GCMC), and kinetic Monte Carlo (KMC) simulations.

Based on a molecular-level description of the i-PI membrane, MD and GCMC simulations are used to generate a lattice model of the pore structure and the local electrostatic environment relevant to the diffusing gas molecules (CO_2/CH_4). This discretized membrane information is used to populate a comprehensive KMC model that can be used to make predictions of gas separation performance on experimentally-relevant time scales. Using this KMC model, we can predict permeability, selectivity, and other transport behavior and connect these metrics directly to the ongoing experimental synthesis work.

The performance of neat i-PI systems is evaluated, as well as composite structures containing both i-PIs and various ionic liquids (ILs). The i-PI+IL composites are based on combinations of 1-n-butyl-3-methylimidazolium ([C_4mim^+]) cations with three common molecular anions: (bis(trifluoromethylsulfonyl)imide ([Tf_2N^-]), tetrafluoroborate ([BF_4^-]), and hexafluorophosphate ([PF_6^-]). It is found that 50 mol% IL inclusion can increase CO_2/CH_4 selectivity by 16% in [BF_4^-]-based materials and by 36% in [PF_6^-]-based materials from mixtures of 5% CO_2 / 95% CH_4. While the [BF_4^-]-based system shows higher CO_2/CH_4 selectivity, the [Tf_2N^-]-based system shows higher CO_2/N_2 gas selectivity. Overall, we find that some of the commonly-used design equations for membrane performance provide inconsistent predictions, as compared to our multi-scale simulation analysis that can capture important molecular-level details (pore blocking, neighbor interactions, pore percolation, etc.).

Copyright © American Institute of Chemical Engineers. All rights reserved.

560732 Computational Design of Physical Solvents for Gas Separation

Recently, we have developed a computational approach [1,2] through the integration of the NIST experimental database, in-house computational database, in-house Monte Carlo and molecular dynamics simulations to screen physical solvents for CO_2 pre-combustion capture from the fuel gas, which mainly contains CO_2, H_2, and H_2O. About 23,000 organic compounds contained in the NIST database were screened and one especially promising compound was identified. Both the lab scale (solvent volume ~ 1 liter) and the recent preliminary bench scale (solvent volume ~ 60 liters) experimental data show that this solvent performs better than the commercial Selexol solvent for CO_2 pre-combustion capture application.

In this talk, we will discuss our recent computational driven efforts to design physical solvents with large CO_2 solubility along with other favorable properties, such as high CO_2/H_2 solubility selectivity, high hydrophobicity, low viscosity, low vapor pressure, and low tendency to foam. From the simple group contribution method, several functional groups with a large CO_2 solubility contribution were identified. Both experiments and simulations show that when the number of -CH_2- groups increase, CO_2 solubility decreases. This is due to the following two reasons. When the number of -CH_2- groups increase, the functional group concentration decreases. This leads to a smaller number of functional groups in the same volume of solvent, which in turn decreases CO_2 solubility. In addition, our simulations show that when the number of -CH_2- groups increase the solvent free volume fraction will decrease, which also decreases CO_2 loading. Furthermore, both our simulations and experimental data show that simply adding more functional groups to a solvent molecule could significantly decrease instead of increasing CO_2 solubility. When more functional groups were added to a solvent molecule, the solvent free volume fraction significantly decreases partly due to stronger solvent-solvent interactions, which in turn decrease CO_2 solubility.

Finally, our most recent results from computational screening of the NIST solvent database will be presented, such as H_2O, H_2S, H_2, and CO_2 interactions with all functional groups. It was found that H_2O interactions with all functional groups are stronger than CO_2 interactions with the same functional groups, and the two sets of interactions show a strong linear positive correlation coefficient of 0.88. New simulated and experimental CO_2 solubility data will also be compared with the results obtained from our previous machine leaning model [2].

References:

1. Wei Shi, Robert L. Thompson, Megan K. Macala, Kevin Resnik, Janice A. Steckel, Nicholas S. Siefert, and David P. Hopkinson, "**Molecular Simulations of CO_2 and H_2 Solubility, CO_2 Diffusivity, and Solvent Viscosity at 298 K for 27 Commercially Available Physical Solvents**", Invited paper to the *J. Chem. Eng. Data*, DOI: 10.1021/acs.jced.8b01228, 2019
2. Wei Shi, Robert L. Thompson, Megan K. Macala, S. Tiwari, Kevin Resnik, Janice A. Steckel, Nicholas Siefert, David P. Hopkinson, "**Integration of Data Mining, Molecular Modeling, and Machine Learning to Screen Physical Solvents for Gas Separation**", 2018 AIChE Annual meeting, David L. Lawrence Convention Center, Pittsburgh, PA, Oct. 28-Nov.2, 2018

Copyright © American Institute of Chemical Engineers. All rights reserved.

571114 First-Principles Computational Study of the Adsorption Mechanism for $CH_3I(g)$ Removal

We study removal of gas-phase organic methyl iodide (CH_3I) from an ambient environment via adsorption onto organic additives, triethylenediamine (TEDA) and quinuclidine (QD), and transition metal impregnants. First principles density functional theory (DFT) calculations and ab-initio molecular dynamics (AIMD) simulations were extensively utilized to understand the underlying mechanism for the chemical reaction of CH_3I. Our results suggest that the adsorption energy of CH_3I shows substantial heterogeneity depending on the adsorption site, porosity of the AC, and humidity. It is observed that the CH_3I dissociative chemisorption is largely influenced by the adsorption site. Organic additives show difference in adsorption performance with their basicity, and transition metals show different adsorption performance according to their bonding characteristics towards iodine. Most importantly, it is clearly shown that humidity plays a critical role in decreasing the adsorption efficiency, but for assisting the adsorption for QD. Our computational study can help to open new routes to design highly efficient materials for removing environmentally and biologically hazardous materials, for example radioactive iodine gas emitted following accidents at a nuclear power plant. We also propose innovative and insightful map to guide sorting out the best metal adsorbents and impregnants for dramatic improvement of the adsorptive removal of the radioactive iodine gas.

Copyright © American Institute of Chemical Engineers. All rights reserved.

577308 Binding Mechanism of Titania Binding Peptide (TBP) on Titania from MD Simulations

Titanium (Ti) alloys are used in dental and bone implants due to their mechanical strength and biocompatibility. Solid binding peptides have been studied extensively in order to impart bioactive properties to implants, and aid in the process of osseointegration (structural and functional integration of the surface of an external implant with the living bone). In order to boost biocompatibility of Ti implants, titanium binding peptide (TBP), a hexapeptide motif (RKLPDA) that electrostatically binds to Ti surface has been extensively studied, however, the binding mechanism of this peptide is unclear. In this study, we employ molecular dynamics simulations enhanced by replica exchange metadynamics to investigate the binding of TBP on titania. We calculate binding free energy and find that it is in good agreement with experimental dissociation constant. We perform cluster analysis to investigate the different binding poses of TBP on the surface and see that binding occurs through a few conformers. Finally, we compare these structures with that obtained from ssNMR, and find that there is a good match between peptide conformations obtained though experiments and simulations.

Copyright © American Institute of Chemical Engineers. All rights reserved.

569101 Rapid Prediction of Adsorption in Hyper-Crosslinked Polymers: Expanding the Chemical Space for Adsorption and Separation

Microporous materials, including ordered and amorphous structures, have been widely used in chemical separations. Hyper-crosslinked polymers show great potential for catalysis, gas storage and separations due to their excellent physical, thermal and chemical stability. Molecular simulations for HCPs can predict adsorption properties ahead of experiments, but the computational burden of doing so is large when aiming to consider a diverse range of molecular species that may be relevant in practical chemical separations. Our study aims to provide a feasible rapid approach towards exploring adsorption space for molecular separations in HCPs. We obtained the full adsorption isotherms of 216 molecule-HCP pairs at 300 K using GCMC calculations by investigating 9 types of HCPs and 24 distinct adsorbing molecules. Using this data, we developed efficient models to predict adsorption isotherms that can be extended for prediction adsorption properties of larger set of molecules and a variety of HCPs based on calculated Henry's constants, liquid phase densities of molecules and helium void fraction of HCPs.

Copyright © American Institute of Chemical Engineers. All rights reserved.

569486 Effects of Topology and Pore Size of Metal-Organic Frameworks on Their Adsorption and Transport Properties: Insights from Molecular Simulations

Metal-organic frameworks (MOFs), which have high crystallinity, large surface areas, well-defined porosities and tunable chemical environments, are emerging as next-generation materials for a broad range of applications related to adsorption, separation and catalysis. The structural features of MOFs, such as topology and pore sizes, can be controlled in a systematic way, which allows precise design for specific applications. In this work, we studied the effects of MOF topology and pore sizes on the adsorption and transport properties via atomistic molecular simulations and established structure-property relationships as design rules. First, we studied the adsorption behavior of ethanol and alkanes in a series of MOFs to explore the potential of MOFs as adsorbent materials in adsorption-based refrigeration systems, which are a promising energy-efficient alternative to traditional compression-based systems. Our results showed that certain MOFs can provide largely improved working capacities over traditional adsorbents and there exists a structure-property relationship between the MOF pore size and the working capacity of the system for each specific working condition. Molecular-level insights on the microscopic adsorption mechanism were also gained from the simulations. Next, we investigated the diffusion of alkanes in MOFs to understand how the structural parameters determine the diffusivity. We looked at a series of hierarchical MOFs that contain both micropores and mesopores with the same topology but varying pore sizes. In general, the introduction of mesopores in hierarchical porous materials was believed to enhance the mass transfer. However, from molecular simulations, we obtained a non-monotonic diffusivity-loading relationship in which the diffusivity first decreases then increases. From detailed analysis of the simulation results, we concluded that the diffusivity at different stages of loading is controlled by different structural factors—the size of the micropores and the mesopores determine the diffusivity at low and high loadings, respectively. We then designed another MOF with that topology and its micropores have the same size as in the MFI zeolite. The diffusivity-loading relationship of n-hexane in that MOF coincides with that in the hierarchical self-pillared pentasil (SPP) zeolite assembled by MFI units, which verified our conclusion. Together, our results shed light on the complex microscopic adsorption and transport behavior in MOFs and may guide the future design of functional porous materials.

Copyright © American Institute of Chemical Engineers. All rights reserved.

570634 Molecular Dynamic Simulation of Protein Devices: Pegylated Proteins and Protein Microarrays

Protein based technological devices have great potential to change how we harness the power of biology. Two examples of such protein devices are PEGylated proteins and protein microarrays. PEGylation is a process primarily used in pharmaceutical therapeutics that covalently bonds polyethylene glycol (PEG) polymer chains onto the protein's primary structure with the propose of reducing the immunogenicity of the drug by slowing renal filtration thus increasing protein's efficiency in the body. Protein microarrays, or proteins that have been covalently tethered to a solid substrate, have potentially transformative applications for detection assays in fields as diverse as proteomics, national defense and biocatalysis.

Creating the aforementioned proteins-based devices requires placing the protein in a situation different than its naturally-occurring environment, and functionalization often renders the molecule inactive. However, at certain sites, functionalization has little effect on stability and can even enhance protein activity. Because the current experimental methodology requires a laborious and expensive guess-and-check approach, being able to predict good attachment site locations that result in stabilization would greatly reduce experimental costs and facilitate rapid development of next-generation devices.

Recently, we developed a coarse-grain molecular simulation method that utilizes MBAR machine learning algorithms to accurately predict the full range of stabilizing to destabilizing attachment sites. In this presentation, we explore an in-depth analysis of ß-lactimase (penicillinase), and how PEGylation and microarray tethering affects this protein's stability.

The results of an in silco screen of all experimentally-relevant surface attachment sites show that for both protein devices, device attachment site strongly affects thermostability, and experiments confirm these results. Moreover, by screening the same protein for both PEGylated proteins and protein microarrys, we show that optimal site locations for one protein functionalization approach is not predictive of the other.

Additional research in PEGylation has yielded new Go-model coarse-grain parameters for polyethylene glycol. Both a purely-repulsive and an attractive force field were used to asses PEG effects on protein stability. The results from these simulations indicate that PEG/protein structures formed in simulation contradict previous theory that PEG stability primarily derives from wrapping around the protein and instead the PEG polymer is attracted to itself and remains removed from the protein body.

Taken as a whole, these results reveal a highly-detailed view of protein-PEG and protein-surface biophysics and reveal the synergistic nature of our computational/experimental approach for the in silico rational design of protein-based devices.

Copyright © American Institute of Chemical Engineers. All rights reserved.

577742 A Twenty-Five Year Perspective on Differences and Possibilities in the Practical Adoption of Molecular Simulations Vs Computational Quantum Chemistry

Computational quantum chemistry has become a practical tool, widely for generating thermochemistry and electronic properties of materials. Molecular simulation has been used selectively in industry, largely for qualitative insights.

I will explore and contrast applications over the past 25 years since AIChE hosted a plenary session for six invited industrial speakers that Paul Mathias, Ken Cox, and I organized for the 1994 San Francisco Annual Meeting (see this link for more pre-CoMSEF details). Additional applications and comparisons are available in the 2002 report *International Comparative Study on Applying Molecular and Materials Modeling*, the 2009 report *International Assessment of Research and Development in Simulation-Based Engineering and Science*, and presentations at this meeting in the CoMSEF session "Industrial Applications of Computational Chemistry and Molecular Simulation."

These experiences suggest that limited routine use of molecular simulations can be understood better through the three elements of cyberinfrastructure - hardware, software, and people:

- Many-electron, many-atom, many-molecule, and many-mer modeling efforts have benefitted greatly from **hardware** advances in computing speed, in memory and storage, and in visualization.
- **Software** packages for computational quantum chemistry have been deployed widely and have high ease-of-use; most are commercial codes although some are open-source. In molecular simulations, excellent software and force fields have long been publicly available.
- Perhaps the weakest aspect has been in **people**. As Ken Cox once noted, the people goals for software have shifted from trying just to be "foolproof and user-friendly" to needing to be "fool-friendly and user-proof." Many people want deliverables quickly but resist learning enough to generate them. Use of molecular simulation still seems limited to experts, and results are often qualitative rather than quantitative. In contrast, computational quantum chemistry has become an easily used tool for generating certain properties accurately with useful precision.

For molecular simulations, achieving broader use will require software capabilities that easily and reliably deliver quantitative, needed results of specific value. I will remark on several trends that promise to help, including advances in raw and specialized computing power; computational material science; Reactive Molecular Dynamics; Energy-Minimization Multi-Scale methods; and data science and machine learning.

Copyright © American Institute of Chemical Engineers. All rights reserved.

569331 *Freud*: Powerful Particle Simulation Analysis in Python

The evolution of increasingly sophisticated particle simulations toolkits has partially shifted the bottleneck from simulation to analysis. The *freud* library aims to ameliorate this bottleneck by providing a simple, NumPy-based Python interface to efficient, highly parallel analysis methods written in C++. The core architecture of *freud* is designed around classes encapsulating the functionality for efficiently finding neighbors in periodic simulation boxes. Analyses in *freud* are generally built around these neighbor-finding facilities, which, in conjunction with the use of NumPy arrays as the primary data structure, enables easy application of any analyses to a particular simulated system. The package also provides access to these core features in Python to facilitate fast-paced development and testing of novel analyses, encouraging a usage pattern where methods are developed and validated in Python before translation into fast C++ code. By completely eschewing any dependence on specific file formats or data structures, *freud* requires minimal configuration to integrate into workflows involving almost any simulation toolkit, making it a powerful tool that can be used to analyze a wide array of molecular simulations.

Copyright © American Institute of Chemical Engineers. All rights reserved.

576970 Molecular Simulation and Design Framework (MoSDef) for Transparent, Reproducible, Usable By Others, Extensible Simulations (TRUE)

Reproducibility in scientific research has become a prominent issue [1]. Computational scientists are grappling with the central question: How can a computational study be performed in such a way that it can be replicated by others? To this end, we discuss the Molecular Simulation and Design Framework (MoSDeF) [2], a suite of Python-tools designed to facilitate reproducible molecular simulations, with an emphasis on the ability to perform large scale screening of structural/chemical space. The core tool, mBuild [3,4], is used to generate configurations of chemical systems of interest. mBuild leverages the concept of generative modeling, where complex systems can be constructed from smaller, interchangeable components, allowing for simple parameterization of structures. Repetitive structures like polymers, crystals, spherical or planar tiling patterns can be expressed declaratively. These features of mBuild allow users to programmatically vary parameters for a family of systems (e.g., polymer chain length) as well as vary interchangeable components (polymer subunits, crystal basis particles, etc) with minimal changes to the underlying code. A plugin architecture has been developed that allows the external development and dissemination of more specialized initialization routines that use the mBuild library, without modification of the mBuild library itself. mBuild integrates with Foyer [5,6,7], a tool for defining force field parameter usage and applying force fields to molecular systems (i.e., atom-typing). Foyer provides a force field and simulation engine agnostic method for defining parameter usage that relies upon SMARTS8 based annotations of chemical context and overrides statements to set rule precedence. As such, these annotations provide both human and machine readable documentation of parameter usage, reducing ambiguities as to how parameters should be used and removing rule order as a source of error. By separating rules from the code to evaluate them, force field files can be constructed, modified, version controlled, and easily disseminated. Similarly, a plugin-environment has been developed for Foyer force fields, that also includes tools for validation and verification. We have coupled this framework with the workflow management tool Signac-flow [9] enabling MoSDeF to perform large scale parameter screening, capturing the exact MoSDeF procedures, inputs, and other relevant metadata. Combined, these tools provide a framework for performing simulations that are Transparent, Reproducible, Usable by others, and Extensible (TRUE).

References

1. Baker, M. 1,500 Scientists Lift the Lid on Reproducibility. *Nature* **2016**,*533*(7604), 452–454.
2. MoSDeF [Online]. Available: https://github.com/mosdef-hub.
3. C. Klein, J. Sallai, T. J. Jones, C. R. Iacovella, C. McCabe, and P. T. Cummings, “A Hierarchical, Component Based Approach to Screening Properties of Soft Matter”, Foundations of Molecular Modeling and Simulation, **2016**, pp. 79-92.
4. mBuild [Online]. Available: https://github.com/mosdef-hub/mbuild.
5. Iacovella, C. R.; Sallai, J.; Klein, C.; Ma, T. “In Idea Paper: Development of a Software Framework for Formalizing Forcefield Atom-Typing for Molecular Simulation”, 4th Workshop on Sustainable Software for Science: Practice and Experiences (WSSSPE4), **2016**.
6. Foyer [Online]. Available: https://github.com/mosdef-hub/foyer.
7. Klein, C., Summers, A.Z., Thompson, M.W., Gilmer, J., McCabe, C., Cummings, P.T., Sallai, J. and Iacovella, C.R., 2018. Formalizing Atom-typing and the Dissemination of Force Fields with Foyer. *arXiv preprint arXiv:1812.06779*.
8. SMARTS [Online]. Available: http://www.daylight.com/dayhtml/doc/theory/theory.smarts.html
9. Signac-flow [Online]. Available: https://bitbucket.org/glotzer/signac-flow.

Copyright © American Institute of Chemical Engineers. All rights reserved.

575245 Large-Scale Nonequilibrium Molecular Simulations of Self-Assembly and Transport Processes

Due to the explosion of computational resources and co-designed efficient software packages in the last decade, molecular simulation has become a valuable tool for interrogating structure and dynamics with microscopic resolution and to rapidly explore parametric design spaces. However, the simulation of many important nonequilibrium assembly and transport processes at the mesoscale often still poses a formidable challenge. In this talk, I will focus on barriers to the application of nonequilibrium molecular simulations using evaporation-induced colloidal crystallization as a case study. I will highlight key challenges that were encountered and overcome: (1) needing to rethink standard simulation algorithms, (2) understanding and developing appropriate coarse-grained models, and (3) analyzing nonequilibrium self-assembly processes from limited simulation data. This case study demonstrates that although tangible progress has been made toward applying molecular simulations to model nonequilibrium problems, significant technical and conceptual advances are still required in future to make them routine. Toward this end, we have released all of our modeling tools as freely available open-source software.

Copyright © American Institute of Chemical Engineers. All rights reserved.

577843 Multirefpredict: An Automated Workflow for Method Selection of First Principles Calculations on Transition Metal Chemistry

Accurate prediction of electronic properties of open-shell transition metal complexes is essential for materials design and catalysis. However, transition metal complexes are notoriously difficult to study accurately with approximate Density Functional Theory (DFT). A crucial step to obtain accurate results for transition metal containing systems is to choose between single-reference and multi-reference based methods. Here we develop a python module, MultirefPredict, which is a high-level cross-platform workflow that calculates widely used multi-reference diagnostics for any given molecular system, without users handling the input or output of quantum chemistry packages. The backend quantum chemistry packages of MutlrefPredict supports several quantum chemistry packages based on both DFT and wavefunction based method, with calculation done on both CPUs and GPUs. We are working on incorporating this module into molSimplify, an automated, open source toolkit that enables seamless generation of candidate inorganic molecule structures, preparation and execution of electronic structure calculations developed in our group. The updated version of molSimplify will be able to recommend appropriate electronic structure methods for each automatically generated inorganic complex structure. This workflow is anticipated to make transition metal chemistry simulation more automated and transparent.

Copyright © American Institute of Chemical Engineers. All rights reserved.

575523 Recent Developments in the *Signac* Data Management Framework

Computational resources for high-throughput data generation offer incredible potential for accelerating scientific discovery, especially if used in conjunction with well-managed computational workflows. The open-source *signac* data management framework enables researchers to maintain well-formed, reusable data spaces from early exploration through production runs on supercomputers. This is achieved through a transparent data and workflow model, as well as a simple and unobtrusive programmatic interface. The framework is application-agnostic, and has been applied in molecular simulations, quantum chemistry, photonics, computational fluid dynamics, machine learning, graph mining, and even organizing experimental data. Recently, the framework has been significantly extended. Among the new data management features are HDF5 integration for rapid access to large numerical data arrays; tools to import and export data spaces for long-term archival and publication of data sets; and increased integration with the scientific Python ecosystem to support easy export to pandas data frames or visualization in Jupyter notebooks. Furthermore, workflow automation has been expanded to support an increased set of supercomputers and allow more complex operations. We show examples of recent scientific applications that demonstrate the efficacy and versatility of *signac* across a range of research domains.

Copyright © American Institute of Chemical Engineers. All rights reserved.

580547 Introduction to the Eleventh Industrial Fluid Properties Simulation Challenge: Temperature-Dependence at High Pressure for the Viscosity of Diphenylethane

This talk will provide background information and industrial relevance for the 11th challenge problem.

Copyright © American Institute of Chemical Engineers. All rights reserved.

580549 Results of the Eleventh Industrial Fluid Properties Simulation Challenge

This talk will announce the results of the 11th challenge.

Copyright © American Institute of Chemical Engineers. All rights reserved.

561024 Grand Canonical Inverse Design of Multicomponent Colloidal Assemblies

We present an inverse design methodology that leverages Machine Learning and a recently developed approach for systematically enumerating crystalline ground states of two-dimensional multicomponent materials. The latter allows us to generate phase diagrams for multicomponent colloidal mixtures which govern their equilibrium self-assembly, while the former enables these diagrams to be optimized to produce a desired target structure. Importantly, this scheme enables inverse design which explicitly accounts for the possibility of phase separation. For k-component systems at fixed temperature and pressure Gibbs' phase rule stipulates there may be as many as k coexisting phases. Conventional inverse design methods typically optimize pairwise interactions in a canonical (NVT) system to yield a Hamiltonian that will best assemble into a desired structure at that same composition. While powerful, the apparent complexity of a chosen target tends to be reflected in the resulting pairwise interactions, making these optimized systems difficult or impossible to realize experimentally. We posit that one way to produce more realizable systems is to rely on mixtures of components with simpler interactions, rather than on a single component with a more complicated interaction. For single component systems only a single (ideally, the target) phase will be thermodynamically stable; however, there is no such guarantee for multicomponent systems. In multicomponent systems containing arbitrarily complex potentials it may be reasonable to expect that the optimized set of interactions will not cause the system to phase separate, however, as simplifications are introduced to make these systems experimentally tractable, this assumption becomes less reasonable. Using our approach, which explicitly accounts for the possibility of phase separation, we optimize mixtures with simple, Lennard-Jones-like potentials to produce complex, open crystals and other lattices with broken symmetries to demonstrate how complex structures can be stabilized in multicomponent systems with very simple pairwise interactions.

Copyright © American Institute of Chemical Engineers. All rights reserved.

570346 Self-Assembly and Phase Behavior of Mixed Patchy Colloids with Any Bonding Site Geometry: Theory and Simulation

Patchy colloids and associating fluids have attracted continued interest due to the interesting phase behavior and self-assembly in solution. The ability to fabricate patchy colloids with multiple attractive surface patches of different number, size, shape, and relative location makes patchy colloids a good candidate as building blocks to form complex advanced materials. However, a theory that clearly relates the self-assembled structures that form based on the anisotropic interactions has been missing. Although Wertheim's theory in the form of the SAFT model is widely used to predict self-assembly and phase behavior in solution, SAFT does not include multibody correlations necessary to model any geometry of association site or sites that can form multiple bonds. We have recently developed a new theory for associating colloids that naturally incorporates multibody correlations based on a cluster distribution approach due to Bansal, Asthagiri, Marshall, and Chapman (BAMC). In this presentation, we demonstrate that the cluster distribution theory can accurately predict the thermodynamic properties and phase behavior of binary system consisting of anisotropic particles with any geometry of bonding site. In particular, we consider self-assembly of Janus particles, Saturn rings, and Ternary particles mixed with solvent colloids that have two directional patchy sites. Good agreement between theoretical predictions and molecular simulation is shown for self-assembly, thermodynamic properties and phase behavior in this system. Such interesting behavior as low density gels or empty liquids is predicted.

Copyright © American Institute of Chemical Engineers. All rights reserved.

565142 A Theoretical Study of Dynamic Governing Equations for Binary Colloidal Systems Combined Bbgky Hierarchy and Maximum Path Information Entropy Principle

A Theoretical Study of Dynamic Governing Equations for Binary Colloidal Systems Combined BBGKY Hierarchy and Maximum Path Information Entropy Principle

Teng Zhao[1], Shuangliang Zhao[1, 1] and Honglai Liu[2, *]

[1] State Key laboratory of Chemical Engineering and School of Chemical Engineering, East China University of Science and Technology, Shanghai, 200237, China

[2] School of Chemistry and Molecular Engineering, East China University of Science and Technology, Shanghai, 200237, China

Abstract

Most of the existing non-equilibrium theories based on the local-equilibrium assumption have difficulty in applying to far-from-equilibrium conditions. Herein, we conduct a theoretical research on non-equilibrium systems that composed of binary colloidal mixtures with a combination of Bogoliubove-Born-Green-Kirkwood-Yvon (BBGKY) hierarchy equation and Maximum Path Information Entropy principle. A set of coupled dynamical governing equations derived from the first BBGKY correlation function. The dynamic variables that characterize hydrodynamic properties of the system include local density profile, local momentum and local kinetic energy abided by the key thoughts of *macro-reproducibility*. After that, we apply principle of Maximum Path Information Entropy to give the least-biased estimation for probability distribution function subject to dynamic constraints and normalization condition. It is noteworthy that all dynamic variables are naturally *path-dependent*. With the help of this statistical inference method, it is provable that the correlative time evolution equations are self-consistent. Moreover, the dynamic equation is consistent with that in the continuous limit with appropriate assumptions. This work provides a new theoretical computational method for dealing with non-equilibrium transportation phenomena.

Keywords: Non-equilibrium; BBGKY hierarchy; Maximum Path Information Entropy principle; Dynamical governing equations; Self-consistent

[1] To whom correspondence should be addressed: szhao@ecust.edu.cn (SZ); hlliu@ecust.edu.cn (HL)

Copyright © American Institute of Chemical Engineers. All rights reserved.

574333 Inverse Design for Self-Assembly of Materials with Targeted Mechanical Properties

Inverse design is a promising yet challenging approach to develop new materials. To create materials with new properties, several inverse approaches have been proposed to design building blocks for target structures. However, the relationship between a material's structure and its properties is often unknown for novel materials. This calls for the development of inverse approaches that can directly target material properties without specific structural constraints. Here, we present an inverse-design approach for particles with interacting potentials to self-assemble crystal structures with targeted mechanical properties. We do so through a novel, molecular-dynamics implementation of the 'digital alchemy' inverse-design approach that extends statistical ensembles to include variation in 'alchemical' degrees of freedom describing particle attributes. We give examples in which model particles interact via isotropic pair potentials that are designed to yield structures with a desired bulk modulus. Our results demonstrate that we can directly target mechanical properties via inverse materials design, and our algorithm can be generalized to other properties.

Copyright © American Institute of Chemical Engineers. All rights reserved.

574259 Solid–Solid Phase Transitions in the NaCl–KCl System

We study the solid-to-solid phase transition in the NaCl–KCl system using molecular simulations. Below around 500°C, this system forms a solid solution with a pronounced miscibility gap between a sodium-rich and a potassium-rich phase [1]. We calculate the phase diagram in this region using the semi-grandcanonical Widom method, which averages over virtual particle transmutations at a fixed system composition [2]. By employing a newly modified version of a Fumi-Tosi style force field [3], we are able to get a remarkable degree of agreement with the experimental phase diagram. Furthermore, we calculate the thermodynamic driving force for nucleation and growth within the coexistence region and extrapolate the location of the critical point.

[1] D. Sergeev, D. Kobertz, and M. Müller, Thermochimica Acta **606**, 25 (2015).

[2] P. Sindzingre, G. Ciccotti, C. Massobrio, and D. Frenkel, Chem. Phys. Lett. **136**, 35 (1987).

[3] D. J. Adams and I. R. McDonald, J. Phys. Chem. C **7**, 2761 (1974).

Copyright © American Institute of Chemical Engineers. All rights reserved.

568218 Exploring the Limit of Self-Assembly: High-X Block Oligomers

Block polymer-based templates with strongly incompatible segments and short chain lengths have gathered much attention due to their micro-phase separations and resulting domains with extremely small feature sizes. Nanopatterning using directed self-assembly of such materials is under consideration as a cost-effective route to achieving sub-10 nm feature sizes, targeted for microelectronic devices with faster processing speed and higher storage density.[1] Motivated by further miniaturization of domain size, our recent work utilizing atomistic-scale molecular dynamics simulations identified a class of amphiphilic oligomers capable of forming ordered lamellar and cylindrical morphologies with domain periods as small as 3 nm.[2] We integrated theoretical and experimental work focusing on a sugar-derived amphiphile with an alkyl tail, and found a lamellar morphology induced by thermotropic self-assembly with a 3.5 nm period and 1.1 nm domains. In addition, structural details from simulations are in remarkable agreement with the experimental results.

1. Sinturel, C.; Bates, F. S.; Hillmyer, M. A. High χ-Low N Block Polymer: How Far can We Go? *ACS Macro Lett*. **2015**, *4*, 1044–1050.
2. Chen, Q. P.; Barreda, L.; Oquendo, L. E.; Hillmyer, M. A.; Lodge, T. P.; Siepmann. J. I. Computational Design of High-χ Block Oligomers for Accessing 1 nm Domains. *ACS Nano*, **2018**, *12*, 4351–4361.

Copyright © American Institute of Chemical Engineers. All rights reserved.

564510 Transferable Multiscale Model for Simulating Self-Assembled Skin Lipid Mixtures

The stratum corneum (SC), which is the outermost layer of the epidermis, acts as the primary barrier between the body and environment. The SC is made up of dead corneocyte cells surrounded by a multilamellar matrix of lipid molecules containing over 15 different ceramide classes, cholesterol, and free fatty acids of varying tail lengths. Simplified model systems that contain only a subset of these lipids have been extensively studied as a means of uncovering the role each of the lipids play in the SC. While these experimental studies have led to considerable insight, they are unable to provide an accurate representation of the molecular arrangement of the lipids. Molecular dynamics simulations are a viable means to obtain a 3D representation of these lipid arrangements. While atomistic simulations provide an accurate representation of the molecular-level interactions, they typically begin from preassembled initial structures, for which the final configurations are likely heavily biased by the initial structure due to the low mobility of the lipids. Additionally, atomistic simulations have typically examined bilayer, rather multilayer structures, often due to computational cost, and thus will not accurately capture the structure. Self-assembled structures that start from a randomized initial configuration yield more realistic data as they eliminate this bias, however, the long timescales required for self-assembly and large system sizes needed to study multilamellar structures, necessitates the use of computationally efficient, coarse-grained (CG) rather than atomistic models.

In prior work, CG models and forcefields for ceramide NS [1], cholesterol [*in prep*], free fatty acids [1], and water have been developed [2] using the multistate iterative Boltzmann inversion (MS-IBI) method. MS-IBI optimizes CG forcefields by considering multiple thermodynamic states simultaneously with the goal of developing a CG forcefield with reduced state dependences and thus increased transferability [3]. Here, we demonstrate the transferability of this CG model and forcefield by examining the behavior of systems containing two additional classes of ceramides: ceramide AP and NP. The underlying CG beads and interaction potentials associated with ceramide AP and NP are taken directly from the ceramide NS forcefield, with no additional MS-IBI optimizations performed; new bond/angle parameters were derived when required. Simulations of self-assembled multilayers containing ceramide AP and NP provide close agreement with corresponding experiments [4]. Thus, these self-assembled structures provide clear insight as to the three dimensional structure of these membranes.

1. Moore, T. C., Iacovella, C. R., Leonhard, A. C., Bunge, A. L. & McCabe, C. Molecular dynamics simulations of stratum corneum lipid mixtures: A multiscale perspective. Biochemical and Biophysical Research Communications. 498, (2018).
2. Moore, T. C., Iacovella, C. R. & McCabe, C. Development of a Coarse-Grained Water Forcefield via Multistate Iterative Boltzmann Inversion. Foundations of Molecular Modeling and Simulation. (2016)
3. Moore, T. C., Iacovella, C. R. & McCabe, C. Derivation of coarse-grained potentials via multistate iterative Boltzmann inversion. J. Chem. Phys. 140, (2014).
4. Schmitt, T., Lange, S., Dobner, B., Sonnenberger, S., Hauß, T., & Neubert, R. H. Investigation of a CER [NP]-and [AP]-Based Stratum Corneum Modeling Membrane System: Using Specifically Deuterated CER Together with a Neutron Diffraction Approach. Langmuir. 34, (2017).

Copyright © American Institute of Chemical Engineers. All rights reserved.

AUTHOR INDEX

AUTHOR INDEX

AUTHOR INDEX

AUTHOR INDEX

AUTHOR INDEX

AUTHOR INDEX

AUTHOR INDEX

AIChE
120 Wall Street, FL 23
New York, NY 10005-4020

CURRAN ASSOCIATES INC.
proceedings.com

ISBN 978-1-7138-0542-7
9 781713 805427